Big Bang to Humankind

Flavien Phanzu Mwaka

Big Bang to Humankind

Astronomy 1

ISBN numérique : 979-10-262-9935-6

EAN papier : 9791026299950

Librinova

Courriel : contact@librinova.com
Internet : www.librinova.com

LEGAL DEPOSIT NUMBER /RDC/BNC : MR 3.02003-57107

Courriel : lesversetsdessavants@gmail.com
Internet : https://payhip.com/LESVERSETSDESSAVANTS

The cover : The brilliance of the Big-bang and Saint Thomas Aquinas

FLAVIEN PHANZU MWAKA; in front of the dome of the Observatory of the Baronnies Provençales (France /September 2019)

From the same author in the collection "The Verses of the Scholars".
– Volume I : cosmology according to biblical and ancient models. (French version : ed. Exibook)

First part : Introduction to Pre-Big-bang (éd. Librinova)

– Volume II : The field of the Universal Cosmological Model (English and French version : ed. Librinova)

– Volume III : Pre-Big bang – Big bang - after Big bang (English and French version : ed. Librinova)

– Volume IV : The enigmas of Universe solved in the Universal
Cosmological Model
(French version : ed. Librinova)

– Volume V : From Noah to Barack Obama
(English and French version : ed. Librinova)

From the same author in the collection "ASTROTHEOLOGY".
– Volume III : Big-bang to Humankind (ed. Librinova)

N.B. : online sales and in bookshops

Author's contact : lesversetsdessavants@gmail.com

Contents

DEDICATION

This book is dedicated to Pope Francis, who said that "the Big Bang is an act of love from God". (SudOuest.fr Published 10/29/2014).

"Man is born pure, it is up to society to make him perfect.".

Flavien PHANZU MWAKA

Epigraphy

If reason seems to contradict faith, the problem is not the Divine, but the interpretation of that faith. "*Also, when a truth is scientifically demonstrable, he who knows it only as an opinion, for a reason that is only plausible, does not understand it*".

(Saint Thomas d'Aquin)

Acknowledgment

We thank all those who were surprised to find their names in this book through their various statements in documentaries, conferences and some articles on the net. We apologist to them and reassure them that some of their comments have been quoted only as testimony and evidence.

In the same vein, I would also like to thank Professor GUBARIKA of the National Pedagogical University of the DRC, and Father MBUMBA PHANZU, director of the Pedagogical Research Center, who, in spite of their multiple occupations, have agreed to reread this work and to make amendments.

I also thank Mr. Patrick Kunyimba; director of Exibook, who turned my dreams into reality by agreeing to publish my first book.

Finally, I would like to thank The Bible societies which these do not hold against me making the verses read directly to my readers rather than paraphrasing them or referring them to a book. In science, it's like stopping the engine and stepping out to open the hood. We used the Louis Segond version and the King James version

Flavien PHANZU MWAKA

PREFACE

*

Good research works with undeniable religious, scientific and cultural depth. The field covered is tough and defensible in the light of reliable material evidence.

This text should be published as a trial balloon for readers unfamiliar with cosmological material combined with biblical evidence.

I would like the publication to be partial (3 or 4 volumes) in successive editions and not all at once because of the rigidity of the addressed material.

Once again thank you and good luck!

Prof Gubarika

04.03.2020

Dean of the Faculty of Arts
National Pedagogical University (UPN)/RDC.

Foreword

The biblical scheme of creation is a scheme that is both scientific and theological. If the Bible has transcribed it in a theological way, in the collection of books "Astronomy & Theology", we restore it in a scientific way.

The Bible is not a scientific book, as those who denigrate it often say. But, let us not forget that it contains sixty-six books. Among these books, there are some that are poetic, prophetic and other stories. Each book contains chapters, some of which also require scientific and not theological knowledge only to develop them. The biblical scheme of creation, in which the author is telling us how the universe was created, is indeed part of the chapters that are both theological in making God the creator and scientific in telling us how God created.

No one until now has been able to see this scientific aspect of the biblical scheme of creation. This was due to the limitation of human intelligence to be able to fathom the scientific nature of the scriptures. We have broken through this barrier after ten years of research to scientifically reconstruct this pattern.

This rendition in scientific language of the biblical scheme of creation is called "The Universal Cosmological Model".

I started writing the books in this series. My goal was to satisfy my curiosity by checking whether the scientific chronology corresponded to the order of creation or the appearance of things according to the Bible.

The second goal was to decipher the other verses

that have a connection with science, because, in matters of revelation, what is written is not necessarily what the author wanted to express. Sometimes he adapts his language according to the level of the audience. Now, everyone knows the level of Moses' listeners, almost 15 centuries before Christ. They are the people who had just come out of slavery according to the Bible. In the end, I found myself, without wanting to, embarked on the world of astronomy.

I had no reference at the beginning, nor any elementary knowledge of astronomy, chemistry, physics and biology. I didn't even know where I should land. My curiosity was strongly aroused when I realized that I was the first to discover through my research, that the Big-bang was biblical.

Some words and names may seem quite new to my readers. My style is simply to explain a word in the texts themselves, or to use common words that readers can easily understand, even when it has a limited meaning in astrophysics. Thus, I ask the specialists in this discipline not to be strict about the words I use, but to stick more to my ideas.

This book has just brought down, in theory of course, what astrophysicists call the "Planck's Wall"; a point that was declared impassable in general relativity.

Therefore, all the comments and remarks contained in this book are binding on me. It does not deal with any church doctrine, and I write only for my own pleasure as an autodidact writer and researcher. It is just like an inventor who shares what he has discovered to the world of science.

This book; "The Verses of the Scholars - Volume I", is a book designed for the public. It scientifically restitutes how the universe came out, its evolution based on the biblical scheme of the creation.

General Introduction

Why Astrotheology and the Universal Cosmological Model? First of all, a model is a diagram, a theoretical representation aiming to report a process, the existing relationships between various elements of a system.

(Cf : vinsvignesvignerons.com/Geologie/Une-breve -histoire-de-la-planete- terre).

This book deals with cosmology according to the biblical model where we make the scientific restitution of the biblical scheme of creation and cosmology according to the ancient models. In the ancient models, we have made the scientific restitution of the thoughts of some philosophers who reasoned about the origins of the universe.

These two principles, namely cosmology according to the biblical and ancient models form the Universal Cosmological Model; not an ancient model given their origins, but, they are a new revolution in today's science.

From the analysis of these models, it turned out that all these we are seeing as theories in today's cosmology are only the scientific development of what the Bible has revealed in pictorial and parabolic form.

The authenticity of the Universal Cosmological Model compared to other scientific models in cosmology lies in the fact that it deals with the Pre Big-bang, passing through the Big-bang until the appearance of man on earth and in harmonious and coherent way. Therefore, the Universal Cosmological Model swallows up all the partial cosmological models that scientists have offered us so far.

"Swallowed" means that it captures them and puts each one at its right place in order to allow the scientists to provide us with the coherent diagram.

Indeed, the biblical scheme of creation is part not only of faith, but also of cosmology. Its understanding requires the learning of scientific and theological language. Hence astrotheology or astronomy and theology.

Until then, researchers have speculated about the Pre Big- bang. But, having discovered it in the first cosmological book, i.e. the Bible, and in the writings of the ancient philosophers, with all the great ensembles that characterized this period, thus I became the first godfather of this discovery.

From now on, all the theories on the course from pre-Big-bang to Big-bang, which will not take into account one of these biblical components as restituted in the Universal Cosmological Model, will be incomplete. This will be confirmed after the development of this volume in the other volumes.

This discovery and restitution have just given birth to the science of Astrotheology and the Universal Cosmological Model which is the scientific restitution of the scheme of creation as contained in the book of Biblical Genesis. We will come back to it later.

Before tackling the theme of this book, i.e. from the Big Bang to man, we will first go back, through the first two chapters, to the notions developed in the previous volumes, otherwise readers who have not yet read it may not understand the symbiosis between biblical language and science.

In addition, after finishing the first part of the book "Cosmology according to biblical and ancient models" in the collection "verses of scholars", I had resolved to glance at the ancient knowledge. I've never had a course in philosophy, or even in theology, but what I discovered on glancing there surprised me greatly.

I believed that the great knowledge of cosmology came only from contemporary science, but, what came to my surprise is to find that our conclusion is only based on the mathematical level. Indeed, the workings of the universe were detailed in ancient times, not only by Moses, but some philosophers as well. I had already started to restore the ideas of these philosophers in the previous volumes and we will continue them in this volume.

Chapter I :
The notion of the time and labor in the Bible

One cannot understand the cosmological scheme of the biblical creation without knowing the notion of time and work in the Bible. These preliminary notions are a basis in Astrotheology and we return to them in each publication.

Section 1
Biblical language

The biblical language used by Moses was spoken in such a way to convey the message to his contemporaries. The vocabulary meanings were not as rich as it is today.

We should know that the spiritual language, contained in books was determined as sacred ones, such as the Bible, the Koran, the teachings of Buddha, Confucius, the Mayan writings or other esoteric books relating to spirituality, are very different from literary or everyday language. The same word can have different meanings.

For example, when the Bible speaks about death, it is about the separation of man with God through sin, but it is not a physical death. Thus, in the following sentence "*the wages of sin are death.*" (Rom 6:23), we see the saints dying young and the elderly sinners die around 99 years old. And when the Bible talks about eating, it is the context of the sentence that determines whether it is something taken by mouth or intercourse.

That is to say, some biblical sentences, although pronounced in everyday speech, have a different meaning, compared to the true reality. It is a spiritual language called parable, and often pictorial.

The notion of time in the Bible

The story of Genesis shows us that there exists one God who has created all things instantly, like a magician; that is wrong according to our literal analysis. There is, indeed, the notion of time in the history of creation and this notion is well defined in the same Bible.

In everything that God has done, there has always been the notion of time accompanied by the sentence : "there was evening and there was morning" (Ge 1 : 5-27).

Despite the biblical verse : "*But there is one thing, beloved, which you must not ignore, and that is that before the Lord a day is like a thousand years, and a thousand years are like a day*" (2

P. 3 : 8), no Christian theologian has been able to relate this verse to the days in the pattern of biblical creation, believing that the latter was the work of a word. It took another revelation from God for the Moon couple for enlightenment to emerge based on this verse.

Indeed, thanks to the book "The Divine Principle" written by Rev. Dr. Sun Myung Moon that I have had the explanation about the days of creation, otherwise this manual could not be written.

Throughout the history of biblical creation, there is one phrase that keeps coming back, namely *"there was evening and there was morning"* to mark both the beginning and the end of a creation, or a day according to the Bible.

The end of a day according to the Bible is equivalent to an era according to science ; which can last for millions or even billions of years.

1.1. The meaning of "one day" in the bible

Indeed, considering the books contained in the Bible as a whole by Christians, whose author, God, is like a hero in the shadows, the biblical day as quoted in the first chapter of Genesis is well defined in the same Bible in two verses :

Ps 90 : 4 "*A thousand years to God are like one day, they are like yesterday already gone, like a short hour in the night.*"

2 Pet 3 : 8 " *But there is one thing, beloved, that you must* *not ignore, and that before the Lord a day is like a thousand years, and a thousand years are like a day.* "

There are two very important terms in these verses, namely *"to ignore"* and *"like"*. First of all, we should not ignore the notion of time. Secondly, there is the comparative term *"like"* : *"a day for God is like a thousand years"*. Here, the word *"equal"* has not been used, but rather *"like"*. This simply means that a biblical day can represent as many days, as many years as a cycle or an era can take.

So to say that creation was made in six biblical days has nothing to do with our twenty-four hour days. Rather, it represents six great stages which, as science has discovered, took thousands, millions, even billions of years.

In addition, the sequence of events as narrated in the first chapter of Genesis is the same with what science has discovered so far : The Humankind being the latest creature to appear on the earth.

A striking illustration which shows that these biblical days are not like the ordinary ones, but they included periods, so, it is during the three first days of creation found in the first chapter of Genesis that there was not yet creation (appearance) of luminaries to mark the notion of time. We notice that the luminaries' appearance took place in Ge 1:14-19 during the fourth period of creation.

Gen*1:14 " God said, "Let there be lights in the expanse of heaven to divide the day from the night; let them be for signs to mark times and days and years;*

..." 19. "Evening and morning came : it was the fourth day."

So, in terms of what criteria did they fix the first three days, since the notion of time began on the fourth day? It was fixed according to the outcome of the evolutions that were unfolding in the Universe.

In order to be able to talk about the notion of time,

three ingredients should be taken into account, namely, a star spinning around another and an intelligent person who is able understand to that. The rotation of the star on itself and its translation or revolution around another, thus create the quantifiable notion of time. Now, as we said it above that during the three first days of creation, there was no existence of the stars. So the days of which we speak in the Bible in chapters one and two of Genesis, constitute periods and not, our ordinary days of twenty-four hours.

1.1.1. Evening and morning according to the Principle of creation

Ge *1 : 4 "... Evening passed and morning came ..."*

Why did the Bible use both concepts "evening and morning" instead of referring to other times' expressions used during the day ? Once again, we resort to the Principle of Creation, a branch of the Divine Principle book of Rev. Dr. Sun Myung Moon as follows :

"The concept" day "in the first chapter of Genesis does not refer to the normal twenty-four-hour of a day, as for the expressions" morning "and" evening "they cannot be interpreted according to the ordinary meaning any more. Consider this passage in Ge1 : 5 "... evening passed and morning came : the first day." After a step of God's creation deeds which had lasted one evening and one night, the coming morning or harvest time was considered by God as belonging to the first day time. Still in Ge 1, God commanded : "there was evening and there was morning : the second day ...the third day ... the fourth day ... the fifth day ... the sixth day. (Divine principle, level 4, P.31)

– The ***"evening"*** means, according to the divine revelation contained in the divine Principle of Rev. Dr. Sun Myung Moon to Which I share the same ideas ", the period when God created or ceased to create. The period during which God ceased to create something or an object until the period when the ideal of creation was reached : it is a period of growth. It is represented by *night*, the moment between the evening and morning.

– the **"morning"** represents the period during which the ideal of creation was realized, that is, the period of maturity or achievement ; which allows God to go ahead to another stage. Hence the biblical sentence «Evening passed and morning came

..." (Ge 1 : 4), to mark the beginning and the end of a period. So we have :

God commanded (said) : which represents his will

Evening : palpable embryonic formation of his will, the seeding moment or the origin.

Night : growth of the created thing or evolutionary period.

Morning : the accomplishment moment according to God's final ideal or the harvest time.

You should know that each of the three steps cited above can take a thousandth of a second (inflationary time of the second day) to several thousand, millions or billions of years until the achievement of the ideal that God seeks, and it is represented by a biblical day (2Peter 3 : 8). This has been explained previously as we can still read them :

– **Peter** *3 : 8 But there is one thing, beloved, that you must not ignore, and that is that before the Lord a day is like a thousand years, and a thousand years are like a day*

Let's examine an excerpt from the papal speech showing that the notion of scientific time is also taken into account by religious in relation to the six biblical days which are only periods and that this is not only an interpretation of Rev. Dr. Sun Myung Moon. This is the speech given at the Pontifical Academy (1951), and then at the astronomers (1952) by Pope Pius XII for encouraging Abbot Georges

Lemaitre who searched and stated to have put forward the idea of the "primitive atom" which is going to become the foundation of the Big-bang theory :

"On the other hand, the avid truth spirit rightly insists by asking how material has never been able to come to such simular state, so inconceivable for our experience today, and on seeking what has preceded it... With the same limpid and critical view that the facts are examined and judged, he forsees and recognizes the work of the almighty creator omnipotent inside whose virtue aroused by the powerful Fiat pronounced billions of years ago by the Creative Spirit, was deployed in the universe, calling into existence, in a gesture of generous love, the material overflowing with energy. It seems, in fact, that the science of today, going back millions of centuries, would succeeded to become the witness of this initial Fiat lux, at that moment during which nothing emerges, with the material, an ocean of light and radiations, whereas the particles of the chemical elements separated and assembled into millions of galaxies by themselves. "

1.2. The existence of the notion of "Labour" in the Bible

There exists well and truly the notion of labour in the history of creation and this notion is well defined in the same Bible by the verbs to do and to form. In everything God has created, there has always been the notion of time and work.

The most striking illustration of the notion of time and labour, and which justifies the effort made by God, we see that in the formation of man and woman. (Ge 2:7-8,19).

Gen *2 : 7-8 "The LORD God formed man from the dust of the earth, breathed into his nostrils a breath of life, and man became a living being. 8Then the LORD God planted a garden* *in Eden on the east side and put the man he had formed in it. "*

Gen *2:19 "And the LORD God formed every beast of the field and every fowl of the air out of the earth, and brought them to man to see what he would call them, that every living creature might be called what man would call it."*.

Here, we are in the presence of a God who is the Creator of all, and who did not give himself a rest time.

Gen 2:21 *"Then the LORD God caused a deep sleep to fall upon the man, and he fell asleep; and he took one of his ribs, and closed the flesh in its place.*

This being so, we see that in everything God has done in the biblical scheme of creation, there has been the intervention of time and work. The misunderstandings between religious and scientists concerning these two notions are thus cleared up. Otherwise, how else can we justify this warrior's rest that God gave himself on the seventh day?

Gen *2 : 1 """Thus were the heavens and the earth finished, and all their host. 2 God finished his work which he had**done**on the seventh day, and rested on the seventh day from all his work which he had**done** . 3 God blessed the seventh day, and sanctified it, because on that day he rested from all his work which he had**created and done** . "*

There is a very important word here that the Bible insists on it to point out the concept of labour; it is the verb *"to do"*. He used to repeat it himself insistently three times to erase the image of an effortlessly fulfilling magical word of God. And in Ge 2 : 3, the verbs "*to create*" and "*to do*" are mixed together, to show that this creation is the result of an effort. A reasoning which until then escaped all theologians' minds. Let us read :

Gen*2 : 3 God blessed the seventh day, and sanctified it, because on that day he rested from all his work which he* *had* ***created and done*** . ”

Section 2
Evolution, the Voice of God and the breaking of symmetries

2.1 Evolution

Ge 1:4 "...there was an evening and there was a morning...".
We said that "evening" means the period when God created or ceased to create and added that the period when God ceased to create a thing or object until the ideal of creation is realised, is a period of growth. It is represented by night, the period between evening and morning.
Scientifically, this dark period is known as "evolution". A thing undergoes evolution when it has not yet reached the final form or maturity required.

Compared to the growth that takes place during a lifetime, evolution is a long period of growth that extends over time, through generations, until an organism reaches a stable level in its biological formation. However, it must be pointed out that this stability is not acquired by a chance circumstance, but was already programmed into the genes of the individual.

2.2. God said

We want to come back again to this magical phrase of "*God said*" and "*the thing was*" in the biblical scheme of creation.
When creation is scientifically analyzed to this day, scientists have discovered that there were no more breaks between the time when the first particles appeared or were formed, until the formation of man (Gen 1:2-27) -

even though science sometimes speaks of spontaneous symmetry breaking at certain periods, or accidental mutations. This is why scientists do not believe in the evolutionary predestination of creation

There has been harmonious cyclical evolution in the order of appearance of things. For example, everything that breathes could not appear in the absence of water.

With this in mind, we must ask ourselves whether God really dictated the appearance of things. In the light of the analysis in this book and what science has been able to discover, we see that this was not the case. For, if it were so, we would have a different DNA structure from plants and animals; this is not the case. For this reason, we can say that things have evolved, and no one thing has appeared independently of the others. From the formation of the first atom, to man, everything is linked. And man, who is the last creature, appears as the microcosm of the cosmos, having within him all that is physical, all that is energy and all that is spiritual in the cosmos.

Let us take a concrete case in the evolution of the universe, but under the divine will :

In Ge 1:3 "God said, 'Let there be light'; and there was light.

And "*there was light*"; but this light came into being, not because God said let there be light, but rather it was a consequence of God's spirit (energy) on the waters (original matter) in the abyss (kind of black hole) (Ge 1:2). So it was a phenomenon obeying physical laws. But, since it was part of the pattern of creation as predestined by God, Moses spoke of "God said" before it appeared.

So why did Moses use the expression "*God said*" every time, when we were dealing with evolutionary and physical- chemical phenomena? It was to say that the formation of the universe was under divine control and this was part of God's ideal. So the expression "God says" has nothing to do with a magic phrase and this is what has obscured the biblical scheme of creation in cosmology up to that point.

To get from one evolution to the next in the Bible, Moses used the phrase "God said" to show that it was not a "chance of nature" that was being accomplished.

2.3. The spontaneous breaking of symmetry

Spontaneous symmetry breaking is biblically matched by the terms "God said, God did, God created".

Symmetry implies a conservation of properties, a kind of global stability even within a movement. The breaking of symmetry therefore implies that a system where something that was conserved suddenly ceases to do so and gives rise to an unexpected phenomenon.

The history of creation, for science, is about the spontaneous breaking of symmetries. Moses, in the Bible, divided them into six days, which are part of the six periods of creation. But within each period there are unexplained phenomena. For example, if we take the first day, or the period minus one in science (i.e. not yet recognized), we can ask the following question :

In the verse *"God created the heavens and the earth* (universe)"; how did the physical (the universe) appear from the spiritual? For science, there was a break in symmetry somewhere. The symmetry was a stable, self-preserving sky. And suddenly, a part of the sphere of the heavens, filled with spiritual particles, was transformed in an unexplained way into a physical particle and antiparticle, becoming "an invisible and empty earth" (Septuagint Bible).

So a break in symmetry implies a sudden change in laws; a transition, physicists would say, a revolution for all those who study social change. For science, many other properties of matter, all of them in fact, have arisen within such transitions that mark the history of the universe.

All this is to say that scientists are aware of the miraculous phenomena of "God said, God did, God created" by noticing phenomena with missing links in the universe. They have replaced this with the terms "symmetry breaking, mutational accident" which means the sudden, even spontaneous, transition to a state in which a new parameter appears that did not previously exist. This is confirmed by the following quote :

"Discoveries in physics over the last few decades have led us to attach great importance to the concept of broken symmetry. The evolution of the universe since its birth is seen as a succession of broken symmetries. When it emerged from the Big Bang, the universe was symmetrical and without structure. As it cools, it breaks one symmetry after another, thus allowing the appearance of an increasingly differentiated structure. The phenomenon of life itself naturally takes its place in this picture. Life too is a breaking of symmetry. (Cfr : F. J. Dyson, Infinite in all Directions, Harper and Row, 1988).

As far as creation is concerned, so far scientists have noted discontinuous phenomena, giving way to other phenomena. But they have not yet found a causal effect. In the biblical creation scheme, spontaneous symmetry breaking took place in the places where Moses said "*God said, God did or God created*".

For a more specific example of symmetry breaking in the animal kingdom, it is man and the ape. Indeed, the ape is often considered scientifically to be the ancestor of man and the chimpanzee. However, what we can see is that there was a break in symmetry that caused us to go from the form of an ape to a being with a form that allows it to do anything that comes to its mind.

Biblically, after the creation of the animals, God paused. To finally say : *"Let us create man in our image"*. We can see immediately that what will be created or formed again will be different from the previous one. Between the preceding and the following, scientific analysis has revealed that there is a break in symmetry, sometimes referred to in biology as mutational accidents of the genes.

Chapter II :
The Universal Cosmological Model

Section 1
Basic Notion of the Universal Cosmological Model

1.1. Presentation

The Universal Cosmological Model is the analytical restitution of the biblical scheme of creation. The word "analytical" is defined in philosophy as an inference from principles based on experience and observation (Dicos encarta).

It is thus understood that this restitution has nothing to do with a sermon based on faith. All our analyses are supported by the experiences, observations and conclusions that were made by scholars in different fields. Others will be in the centuries to come.

Despite my research and knowledge, I have not reached a full understanding of this scientific restitution, and I am not yet able to express it perfectly. This consideration concerns especially the Before-Big-bang, because the After-Big-bang is widely analyzed by science.

Therefore, the development of the biblical scheme of creation, as reconstructed and analyzed in this book, is called the "Universal Cosmological Model". Our restitution can be criticized by anyone, as long as he does not touch the word "Bible or biblical model" for respecting the latter name and not to tarnish its sanctity. This is how we have entitled our restitution, which can be criticized, "the Universal Cosmological Model". Thus, in making our scientific commentary on the biblical creation story, we enter directly into the Universal Cosmological Model.

Thus, by making our scientific comments on the biblical creation story, we enter directly into the Universal

Cosmological Model, from Astrotheology.

The Universal Cosmological Model, we have said, is the starting point from which everyone can research and draw comments and conclusions. That is to say that nobody, not even I, has a monopoly on the analysis of the Biblical Model. The latter is the set of all the Models that support the biblical pattern on the creation of the Universe. What I have developed and designed is called "The Universal Cosmological Model". Everyone can be inspired and enriched by this Model.

1.2. The basic statement of the Universal Cosmological Model

Thus, the Universal Cosmological Model, on which all analyses will be based, reads as follows :

> *The Universe was formless and empty. Given, on the one hand, an energy defined as darkness by the Bible in the abyss, and on the other hand, the spirit of God or a wind from God on the original material called "waters", there was the production of a radiation defined as light. From the beginning of the Universe in nothingness until the first light or radiation was produced, this is the period that the Bible calls 'day' (first day), a period that lasted for a time that could not be quantified until then.*
>
> The Bible further specified that this new energy or radiation had an effect on darkness. It came to counteract it, as well as the invisible material that was already at work in the early universe. This is expressed in the sentence "God separated the light from the darkness".

And to show that these two energies would henceforth accompany the physical destiny of the Universe, God called darkness "night" and light or new radiation "day", for lack of the appropriate scientific words at that time. The original material or "waters" according to the biblical term, continuing its warming cycle after the production of light (light be and light was), had a burst, an expansion or an expulsion into the abyss and that is what science calls "Big-bang" and the Bible calls "separation of the waters from the waters. This evolutionary period of material warming (waters) from the production of light or first radiation, to the Big-bang, was described by the Bible as the second day. And then came the third day and the fourth day which marked the beginning of the post-Big Bang period which were periods of the creation of the stars from the same material or waters according to the Bible.

Then came the fifth biblical day, the period that witnessed the appearance of living beings from microscopic underwater beings to birds.

And at the end or on the sixth biblical day, there was the appearance of mammals and of humans. (Gen 1:1-30; Gen 2:1-7).

This constitutes the scientific restitution of the writings of Moses dated since 1500 B.C. This is the basic statement of the *scheme of creation according to the Universal Cosmological Model.*

These are exposed in a new way for astrophysicists, cosmologists, astronomers, theologians, philosophers and my other readers, the Biblical Model on which all the other Models that will support it should be built.

Section 2
The biblical creation scheme

2.1. The stages of the universe's evolution according to the Bible

The verses below make us think of the different sentences about the creation history of the universe according to the Bible, also certified by science.

1. Ge *1 : 1-2 "In the beginning, when God created the cosmos, (the heavens and the earth). The earth* (the universe) *was formless and empty ".*
2. Ge 1 : 2 *"There was darkness in the surface of the abyss* (the first deformation of the fabric of space-time) *in the axis of the universe".*
3. Ge 1 : 2 *"The power of God was moving over the water* (original materials before the Big-bang in the deep abyss)"
4. Ge 1 : 3 *"God commanded, 'Let there be light, and light appeared';* (consequence of the Spirit of God moving over the water or over the original material) (Ge 1 : 2)
5. Ge 1 : *4 "God separated the light from the darkness"*
6. Ge 1 : 7 *"Separating the waters from the water"* or Big bang according to science (1) and expanse creation (2). (Beginning of the after Big-bang)
7. Ge 1 : 9 *"From the waters below* (1) *the dry* (2) (or the earth) *appeared"*
8. Ge 1 : 14 *"From the waters above* (1) *there was creation of lights* (2)"
9. Ge 1:20 "Out of the waters there arose life".
10. Ge 1:24 *"God said : Let the earth bring forth the living creatures of the earth according to their kind ".*
11. Ge 2 : 7 *"And God formed man from the dust of the*

earth, and breathed into his nostrils the breath of life (spirit) and man became a living creature.

12. Ge 2:22 And God formed a woman from the rib which he had taken from the man, and brought her unto the man.

All these phenomena or developments take place with the same original materials, so-called as “water” according to the Bible. They are activated by the same energies, also so-called ”darkness” and ”the spirit of God». These are phenomena that make the Universe isotropic (isotropy : The Universe has the same properties in all directions) and homogeneous (Homogeneity : the physical laws are the same in each place).

This tells us that the Universe does indeed have an origin and a starting point as it is described by the Bible. This is what is reproduced scientifically in this book under the name of ”The Universal Cosmological Model” and which gobbles the Standard Model, while correcting the other models and scenarios imagined by scientists.

Indeed, the development of human models and their conclusion can only find their reason in the “Universal Cosmological Model”.

In addition to its steps, I recommend my readers to follow the first five minutes of the scientific documentary entitled : ”Hubble from Space : The Work of Light” on You tube.

In fact, without the director knowing it, this documentary retraces the scenario of the reappearance of light 380,000 years after the Big Bang. But, it said nothing about the appearance of this light since its beginning. This light, according to the Bible, did well and truly appear before the Big Bang or ”the separation of the water from the waters.” By the way, this scenario is similar to what is described in Ge 1 : 1-3.

Let’s read :

Gen 1:1 In the beginning God created the heavens and the earth. 2 The earth was formless and void, and darkness was over the surface of the deep, and the Spirit of God moved over the waters. 3 God said, "Let there be light. And there was light.

In the above-mentioned documentary, the gravity plays the role of the spirit of God whereas the elementary particles play that of the "biblical waters", then, the concentration of particles creating space-time and making the universe an abyss. That is how the appearance of light came. Here is a proof to support in force the Universal Cosmological Model.

Why "in force"? Because it is a scientific proof which has nothing to do with this book and which was conceived out of any idea written in the Bible. In addition, the director talks about a scenario after Big-bang, whereas for us, it is a scenario similar to what took place before the Big-bang until the emergence of the light which is written as follows : "God recommended : let there be light !" And there was light" (Ge 1 : 3).

We have already explained that the phrase "God says" is to demonstrate that this was part of His will. But scientifically, it was the gravitational result on the original matter known biblically by the phrase : "the Spirit of God moved over the waters" (Gen 1:2) as we can see in the above mentioned.

2.2 Illustrative tableau on biblical and scientific cosmology

BIBLICAL PERIODS	CREATION	KEY WORDS	SCIENCE
SPACE-TIME	Ge 1 :1 In the beginning, God created the heavens and the earth.	-Heavens (Spiritual World) ,- Earth (Physical world)	Before the Big-bang
DAY 1	Ge 1 :2-3 « 2 The earth was formless and empty; there was darkness on the face of the deep, and the Spirit of God was moving over the water. 3 God said: "Let there be light! And the light appeared"	- Empty - Darkness - Abyss - Spirit of God - Waters - Lights	
DAY 2	Ge 1 :6 God said, "Let there be an expanse between the waters, and let it divide the waters from the waters."	- Extent-Water separation	The BIG-BANG
DAY 3	Ge 1 :9-11 God commanded, "Let there be a dome to divide the water which are below the heavens be gathered together in one place, and dryness appear. And so it was.10 God called the dry land earth, and he called the cluster of waters seas. "11 Then God said," Let the earth bring forth greenery "	- Seas - Earth - Herbs (Molecules of life)	- Formation of elements up to galaxies (stars, lights and planets)

BIBLICAL PERIODS	CREATION	KEY WORDS	SCIENCE
SPACE-TIME	Ge 1 :1 In the beginning, God created the heavens and the earth.	-Heavens (Spiritual World) ,- Earth (Physical world)	Before the Big-bang
DAY 1	Ge 1 :2-3 « 2 The earth was formless and empty; there was darkness on the face of the deep, and the Spirit of God was moving over the water. 3 God said: "Let there be light! And the light appeared"	- Empty - Darkness - Abyss - Spirit of God - Waters - Lights	
DAY 2	Ge 1 :6 God said, "Let there be an expanse between the waters, and let it divide the waters from the waters."	- Extent-Water separation	The BIG-BANG
DAY 3	Ge 1 :9-11 God commanded, "Let there be a dome to divide the water which are below the heavens be gathered together in one place, and dryness appear. And so it was.10 God called the dry land earth, and he called the cluster of waters seas. "11 Then God said," Let the earth bring forth greenery "	- Seas - Earth - Herbs (Molecules of life)	- Formation of elements up to galaxies (stars, lights and planets)

Chapter III :
The Pre-Big-bang

Gen 1 : "*1 In the beginning God created the heavens and the earth. 2 The earth was formless and empty, and darkness was over the surface of the deep, and the Spirit of God moved over the waters. 3 God said, "Let there be light. And there was light. 4 God saw that the light was good, and God separated the light from the darkness. 5 God called the light day, and he called the darkness night. So there was evening and there was morning—the first day*".

In the previous volumes, we gave a long account of the biblical creation scheme on its first day, constituting the pre-Big Bang. Here, we go through it briefly before tackling our theme from the pre-Big Bang to man.

One cannot understand the Big-bang phenomenon better if one has not yet understood what happened in the Pre-Big-bang. Being the first to have discovered this sphere, as I said, we have described it in Volume II of the collection "Astronomy & theology". Here we will make only a small summary.

For a little reminder, the Pre Big-bang is a period that stretches from an infinite void, as the Italian Physicist Gabriele Veneziano said, to the Big-bang. Biblically, it is described as a period that spreads from a shapeless and empty Earth signifying the universe, until there is a separation of the waters from the waters or Big-bang.

The two concepts are in harmony, apart from the terms used which seem to differ, namely "shapeless and empty earth" for "infinite emptiness" and "Big-bang" for "the separation of the waters from the waters".

In Volume II, which is entitled "Big-bang to fossil radiation", we scientifically restored all the biblical terms that are part of the structures of the Pre Big-bang and which makes me the first person to be able to really discover the origins of the Universe.

Section 1
The universe is a matter of Astrotheology

It's been a century since science approved that the Universe began as a primitive atom according to the words of the Belgian Abbot Georges Lemaitre. Experiments and observations have enriched his theory giving birth to the standard theory in cosmology. But the question everyone is asking is how did this primitive form of atom come about?

Several answers are being proposed, but the most surprising for everyone is that the real answer is in the Bible in the Bible. This answer was discovered by only one person; the author of this book. Who could believe that the description of the Pre Big-bang was biblical with all the structures without neglecting anything?

Who could also believe that Pope Pius XII, described the origins of the Universe in a scientific way before the Italian Physicist Gabriele Veneziano? Finally, when we take Moses, the author of Genesis, Pope Pius XII, the one who was the first to describe the origins of the Universe, Rev. Dr. Sun Myung Moon, the first to reveal the causes of the functioning of the Universe, we understand that the Universe is a matter of Astrotheology.

Although the pre-Big Bang was the subject of Volume II entitled "Before the Big Bang to Fossil Radiation" in this brief development, we will hear the voice of the religious who had no audience in the scientists' manuals, but who are more influential in convincing people than the latter.

1.1 Moses and the biblical scheme of creation

What are the verses that describe the Pre Big-bang?

Gen 1:1 "*In the beginning God created the heavens and the earth".*

2 The earth was formless and empty : there was darkness over the surface of the deep, and the spirit of God moved over the waters. 3 God said, "Let there be light. And there was light. 4 God saw that the light was good, and God separated the light from the darkness. 5 God called the light day, and he called the darkness night. So there was evening and there was morning—the first day. **(Pre Big-bang).**

6 God said, "Let there be a vault between the waters to separate water from water." 7 So God made the vault and separated the water under the vault from the water above it. And it was so. **(Big-bang)***8 God called the vault "sky." And there was evening, and there was morning— the second day..* **(And, beginning of the after Big-bang)**

1.1.2. The creation of the cosmos

Gen 1 : In the beginning God created the heavens and the earth.

In science, when we speak about the Big-bang, we see the physical sphere which preceded the big-bang. In this verse on the other hand, of which I am the only one in the world to be able to discover what it means, even before the beginning of the physical sphere described as before the Big-bang, there were first the heavens.

Physically, astrophysicists know that the sky as a material element does not even exist. It is only a void. The blue that we see is only a limitation of our sight, the color is caused by the combination of the sun's rays with the gases contained in the atmosphere.

Even Moses who lived 1500 years B.C. knew this more than we do. The rest of us learn it in our universities. Moses - the scholar of scholars - had taught since ancient times that the sky is empty space through this verse : "God called the expanse of heaven" (Gen 1:8). This means that after having separated the waters (Gen 1:7) which in fact represent the elements from the Big-bang, Moses does not say that it is the waters above that form heaven but rather the expanse or space between all material structures.

Moreover, the sky spoken of in the verse "God called the expanse heaven" (Gen 1:8) is physical, i.e. containing the physical energies that are known to astrophysicists. But the heaven in the verse "In the beginning God created the heavens and the earth" (Gen 1:1) is a spiritual heaven. It is the sphere that the believer calls the spiritual world. It is from this world that the first spiritual energies came from, which became denser to become virtual energies, accompanied by the phrase "In the beginning God created the heavens and the earth".

This first earth, is not also the earth which is described in the verse which says "10 God called the dry ground "earth" (Gen 1:10) which sprang from water. The earth which is spoken of in the verse "In the beginning God created the heavens and the earth" (Gen 1:1), was a formless and empty earth (Gen 1:2), that is the Universe at its beginning where there was no material structure, except the manifestations of virtual particles going from collisions, annihilation, creations, assorted by the biblical verse "and the spirit or a wind of God moved over the waters". It is the presence of these virtual particles which distinguished "a shapeless and empty earth" meaning the Universe of the sky or the spiritual world.

As we can see, Moses gave his lesson very well but nobody understood it except Muhammad, the MOON couple, Flavien PHANZU MWAKA and Jesus who said he came from heaven, that is to say from the spiritual world. But not understanding what the word heaven means like all believers today except for the followers of the Moon couple, the Jews said to him "Don't we know your parents and your brothers, how do you say that you came down from heaven? "Like all Christians today, the Jews were referring to the physical heaven where they believe Christ will come down again. Of course this is an allusion, which scholars cannot believe.

I make my discoveries as a versatile self-taught researcher. My research has led me to go further than all other researchers who have spoken about the origins of the Universe in modern physics.

Indeed, having discovered the origins of the Universe in the writings coming from the papyrus and inscribed in the Bible, and being the first to give them their true explanations, I thus become the first person to discover the origins of the Universe in the same way as explorers did. America has always been there with its Native Americans, but when Christoph Columbus discovered it, he became the first to discover America. We don't even know the name of the first Native American.

Likewise, the pattern of the creation of the Universe has always been there, in the Bible. People have written big books on Bible commentaries, saying that they are inspired by the Holy Spirit. But, concerning the biblical scheme of creation, before my books were published, they had all reasoned like kindergarten students. For to bring to life this scheme, once thrown away by the scholars, besides the Holy Spirit, it was also necessary to have knowledge of the Principle of Creation of the Moon couple and notions of cosmology. This is what makes me the first person who discovered the origins of the Universe.

What is the difference between scientists' research and the biblical scheme regarding the origins of the Universe? Scientists do research based on observations and experiments to discover the origins of the Universe. But, the Bible tells how the Universe was created from the void to man, which is making the field of the Universal Cosmological Model. What the scientists will discover, in fact, will only confirm the biblical scheme of creation as depicted in the Universal Cosmological Model.

Until then, science had not yet reached the biblical limit, since it still comes up against Planck's wall and the theory of everything still does not hold up. An honorable mention is given to the Italian scientist physicist Gabriele Veneziano, who, without referring to the Bible, discovered in his research that the universe began in an infinite empty sphere in space, thus unknowingly confirming the biblical verse that says "In the beginning God created the heavens and the earth. The earth was formless and empty" (Gen 1:1-2). This means "In the beginning God created the spiritual and physical worlds. The physical world or Universe was formless and empty" (Gen 1:1-2).

We will come back to the scholars who reasoned about the origins of the universe in other volumes. All this is part of my personal discoveries within the framework of the Universal Cosmological Model.

Ge 1:1 "In the beginning God created the cosmos (the heavens and the earth)".

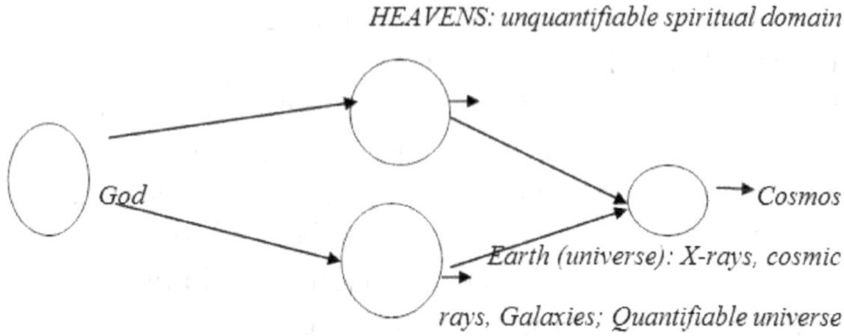

Ge 1 : 2 The earth was shapeless and empty : something that exists but cannot be felt; therefore, without form, i.e. the primitive Universe.

The cosmos, as we have said according to the Universal Cosmological Model, which is in line with the Creation Principle of Rev. Dr. Sun Myung Moon, represents the invisible (spiritual) and visible (Universe) world that the Bible speaks of (the creation of Heaven and Earth). As written previously, the separation of two names in the biblical creation scheme by the conjunction (and), or even by the comma (,), means that they were not created simultaneously.

This separation, as we have said, may have lasted thousands, even millions or billions of years. So when we talk about the creation of Heaven and Earth, the latter meaning the universe, we see that the invisible world (or angelic world, or spirit world) preceded the creation of the universe represented by the earth in the Bible.

The material structures of Pre Big-bang which emerged in a formless and empty earth, meaning the universe, we described and developed them in the first part of this Volume being the subject of a separate book considering its volume.

1.2.The vision of Pope Pius XII

Pope Pius XII was very interested in cosmology and had a very deep sense of the origins of the Universe. Only that, if a religious person tells a truth even in science, it is not taken into consideration by scientists. The words of the clerics are considered simultaneous and do not attract any analysis from the scientists. But, if it is an atheist, it is all the scientists who rush, voting heavy budgets to confirm it with the help of laboratories.

On November 22, 1951, Pius XII delivered the speech "Un'Ora" before the Pontifical Academy. This speech was badly received by the scientific circles which saw in it an obstacle to the scientific progress. Passionate about astronomy, Pius XII wanted to testify to the importance of science. He relays this thesis and evokes, on several occasions, the theory of the primitive atom. He never mentions the name of Lemaître, but the allusion was obvious. (Cf : *Pius XII and the Un'Ora speech - Georges Lemaître, father of the Big Bang Theory - UCL Archives (uclouvain.be)*

When the Belgian Physicist Father Georges Lemaitre expressed his thought of the primitive atom, which will become today the Big-bang theory, he was fought by his contemporaries. He was called upon to introduce religion to science. But, what is the harm if religion says something that science should exploit? That's what I do in my publications.

To come back to our subject, although the primitive atom theory was fought even by the famous Albert Einstein, Pope Pius XII was the first to discover that Father Georges Lemaitre had discovered one of the structures of the Universe. He publicly encouraged it. However, this displeased Father Georges Lemaitre, who felt that the Pope's intrusion was still in danger of proving his detractors right. The Pope was warned not to interfere because his discovery is purely scientific and is not based on the sacred texts. And he will even say that religion no longer interferes with science.

According to some sources, this is where the schism entered between religion and science, and, interpreted on the origins of the Universe had become a matter for astrophysicists. They came up against questions they could not answer and created a god called "the chance of nature".

1.2.1. The origins of the Universe according to Pope Pius XII

What I am talking about is a personal discovery that escapes even the Catholic Church.

Until then, I read the speeches of Pope Pius XII without drawing attention to the words he had spoken. If these words had been pronounced by another person, the Church would have sanctioned it. Here is an excerpt from that speech which also confirms the Universal Cosmological Model. It is the speech which was delivered to the Pontifical Academy (1951) and then to the astronomers (1952) by Pope Pius XII to encourage Father Georges Lemaitre to put forwards the idea of the "primitive atom" which will become the foundation of the Big-bang theory. I would ask astrophysicists, theologians and students of these disciplines to read this very carefully :

"*On the other hand, the truth-seeking mind rightly insists on asking how material could never have reached such a state, so inconceivable for our experience today, and on searching for what preceded it... With the same clear and critical gaze with which it examines and judges the facts, he sees and recognizes the work of the creative Almighty whose virtue, aroused by the powerful Fiat pronounced billions of years ago by the Creator Spirit, has unfolded in the universe, calling for the existence, in a gesture of generous love, of material overflowing with energy. It seems, in truth, that today's science, going back millions of centuries in one fell swoop, has succeeded in witnessing that initial Fiat lux, that moment when out of nothingness arose, with material, an ocean of light and radiation, while the particles of the chemical elements separated and assembled into millions of galaxies.* »

For a conservative Pope, such a speech could not be given, and that is because of the points we are going to develop. But, if he did make it, it shows his open-mindedness to science.

– "*... the creative almighty power whose virtue was aroused by the mighty Fiat pronounced billions of years ago by the Creative Spirit...* " (Pope Pius XII)

– "*It seems, in truth, that today's science, going back millions of centuries in one fell swoop, has managed to make itself the witness of this initial Fiat lux...*". (Pope Pius XII)

Believers are called to conform to the 6 days of creation and the 6,000 years of human history. These quotes from Pope Pius XII contradict this.

When we talked about the notion of time in the first chapter of this book, we said that Rev. Dr. Sun Myung Moon was the first religious to be able to apply time as described in 2 Peter 3:8 to the biblical scheme of creation in his teaching of the Divine Principle. And, there he came to the conclusion that the biblical days are times.

Reading these excerpts from Pope Pius XII's speeches, we see that even the church has taken into consideration that Bible days are periods. We have explained this in Chapter I of this book and in detail in the first part, which is the subject of another book.

And, we concluded by saying that the three stages of the period of the Universe, i.e. the pre-Big-bang, the Big-bang and the post-Big-bang, divided into 7 days in the Bible corresponded scientifically either to a thousandth of a second (inflationary time of the second day) to several thousands, millions or billions of years until the achievement of the ideal that God is looking for, and is represented by a biblical day (2Peter 3:8) as we can still reread it :

2 Peter 3:*8 But there is one thing, beloved, that you must not ignore. It is that, before the Lord, a day is like a thousand years, and a thousand years are like a day.*

I had said in the first part of my book, when a sermon said in church, is published in book form, it becomes a scientific work ready to be analyzed. I have not read the whole of Pope Pius XII's speech delivered on the occasion, but this little excerpt that I discovered on the Web allowed me to decipher that he was also a self-taught cosmologist like me.

Moreover, it was not a sermon, but a speech delivered in front of scientists, even though no one analyzed it. It was like the Moon Couple who in the past brought together many scientists, but nobody had analyzed the Principle of Creation, which alone can lead to the making of the theory of the Whole.

In the speech of Pope Pius XII, we have the following sentence :

"*the creative Almighty whose virtue, aroused by the powerful Fiat pronounced billions of years ago by the Creative Spirit, has spread throughout the universe, calling for the existence, in a gesture of generous love, of material brimming with energy. It seems, in truth, that today's science, going back millions of centuries in one fell swoop, has succeeded in witnessing that initial Fiat lux, that moment when out of nothingness arose, with material, an ocean of light and radiation, while the particles of the chemical elements separated and assembled into millions of galaxies.* "

There are some key words here that you should know how to analyze if you are an enlightened person. For those who have read the first part of this book, the philosopher Empedocles, said that during the period Pre Big-bang (it is I who is clarified by his idea), all the elements were undergoing the contraction by an energy that he described as love. Today we talk about gravitation. And, it is until the Universe becomes condensed that he described Spherus.

In the papal discourse, he evokes the love that was

deployed to bring back into existence material overflowing with energy. Everything arose from nothingness (the earth was formless and empty) as the Italian physicist Gabriele Veneziano later said in the model of the strings. And, material during this pre-Big-bang period was characterized by an ocean of light and radiation, coming from the particle-anti-particle coalition, the difference in rotation of the spheres generating gamma bursts, until the gravitational collapse where the first light germinated : "God said that light be, and, light was".

As we can see Pope Pius XII described the first biblical day, just as the Italian physicist GABRIELE VENEZIANO unknowingly described it. This pre-Big-bang period was concluded with this sentence of Pope Pius XII :

"...as the particles of the chemical elements were separating and assembling into millions of galaxies. "

Separating and assembling are also part of the words that the philosopher Empedocles had to use to mean the Big-bang, the expansion and gravitation of the elements until the formation of galaxies.

Developing all this, we will come to the conclusion that Pope Pius XII was among the first to be able to explain the genesis of the Universe, except that in his time many of the phenomena were still dark to say the right words.

God being a hero in the shadows.

1.3. Rev. Dr. SUN MYUNG MOON (1920 – 2012)

Born in 1920 in North Korea, Rev. Dr. Sun Myung Moon studied electrical engineering in high school. He was at the age of 16, then a fervent believer in the Presbyterian Christian Church, that Jesus appeared to him and told him to complete his mission. At the age of 19, he went to Japan to study engineering at Waseda University.

This training allowed him to explain the origins of divine energy to the functioning of the universe in scientific terms. This is a new revolution in theology and philosophy, and why not in science. His teaching brings together the two disciplines, religion and science, which until then had been separated. And today we have just put a bridge between the two, through Astrotheology.

1.3.1. The Principle of Creation

The Principle of Creation is the branch of the Divine Principle teaching of Rev. Dr. Sun Myung Moon. As the name suggests, it deals with the points related to the causes of the origins and functioning of the universe up to the integral man. He is the only person in the Universe who has argued about the cause of the Universe's functioning. From the religious point of view, we talk about revelation, but in the Universal Cosmological model or in science, we call it predictions.

The Italian scientist Gabriele Veneziano said the :
« universe came from an infinite sphere in the past ». In the Universal Cosmological Model, on the other hand,

we say that the infinite in the past was the spiritual world. Thus, the Universe had its beginning in a physical way, that is to say, putting religion aside, by the densification of spiritual energies transforming into physical energy.

The Rev. Dr. Sun Myung Moon is the only scientist who has mastered how this process of transition took place between the spiritual world and the physical world, that is to say the transformation of the divine spiritual energies into physical virtual energies. This is the subject of his book "The Thought of Unification", taught in the seminars of his church.

To understand this book "Unification thought », one must first have some knowledge of the Creation Principle. That is why we do not even touch it in the Universal Cosmological Model. At least for the time being, until we ourselves can master this philosophical and religious teaching and put it into a language that my readers can understand. However, my knowledge of cosmology has made me the only one who understands that the book "The Thought of Unification" deals with the transitional processes between the spiritual and the physical world.

1.3.2. Speech of Rev. Dr. Sun Myung Moon on the Universe

We said that the Rev. Dr. Sun Myung Moon is the only religious who has contributed a lot to explain the Universe, including the reason of its creation. To develop all these thoughts to the matter, could make the object of a book. Thus, I prefer to put only some extracts of his writings and sometimes without comments so that astrophysicists and cosmologists judge for themselves.

(Presentation of the Divine Principle ; ed. Culture et Paix, Paris, 2013)

– As the fullness of time approached, God sent a person to earth to solve the fundamental problems of life and the universe. His name is Sun Myung Moon. (P.6)

– Throughout history, fundamental questions about life and the universe have plagued human beings without satisfactory answers. This is because no one has understood the underlying principle by which the universe and humanity were originally created. (P.20 level 4)

– Similarly, the activity of energy serves to form particles, the activity of particles serves to form atoms, the activity of atoms serves to form molecules, the activity of molecules serves to form material, and the activity of material serves to build the universe. What is the purpose of the universe? What is its center? The answer to these two questions is none other than the human being. This is why God, after having created him, gave him the mission to rule the universe. If there were no human being to appreciate the universe, it could be compared to a museum without visitors. (P.38)

– Before creating human beings, God created the natural world by expressing in it some of the aspects of the inner nature and outer form that He had planned for human beings. That is why each individual possesses within himself the totality of the essences of all things. That is why he is called the microcosm of the universe as said by God. (P.46)

– Considering that the account of the creation of the universe, reported in the Bible several millennia ago, overlaps to a large extent with the discoveries of modern scientific research, we see in it a confirmation that this biblical text is a revelation from God. The universe did not suddenly spring into existence without any consideration of time. (P.53)

– The order of creation for all things is found in Genesis

1. God's creation began when God created light out of chaos, nothingness and darkness and required a period of six days before it culminated in the creation of man. But since 2 Pet 3:8 states, "...for one day is as a thousand years before the Lord, and a thousand years as one day," we understand that these days were not actually 24-hour days. What Genesis I tells us is that the universe did not exist instantaneously but was created in six gradual periods. (P.32) In reality, its origin and development occupied a long period. Therefore, the six-day period for the completion of the universe should not be understood as a literal succession of six sunrises and sunsets. Rather, it symbolizes six successive periods in the process of creation.

– The universe, with the human being as its center, was created according to mathematical principles in order to express, as a substantial object partner, the dual characteristics of the invisible God. This is the reason why science, which seeks to discover the external laws governing the universe, progresses in its search with the help of mathematics.

– The whole existing world is marked by a single principle, and the whole universe has the same unity and harmony as an organic body. For this reason, all existing beings, from heavenly bodies to atoms, are related to each other, are ordered and exist according to the polarities of mind and body, inner character and outer form, life and organic body, essence and phenomena, time and space ; and yet all these relative factors are united.P.31

– The story of the creation of the universe in six days can be considered as the story of creation through the principle of positivity and negativity. The development of the universe was thus the continuation of a great symphony that achieved ideal beauty and was played by countless instruments called Sung Sang and Hyung Sang (each with positive and negative aspects). The symphony continues today. (P.37)

– But since the universe is a four-dimensional world in space and time, the fourfold base must be formed within the boundaries of time and the three dimensions of space. (P.60)

– Regardless of its vastness, the universe was created to give man satisfaction and joy. Thus, it was created as man's object. Man is the lord and the whole universe is the being over which he rules. If we compare the relative importance of man and the universe according to the purpose of creation, the value of man is greater than the total value of the whole universe, because the object exists for the subject. (P.68)

– Although today man does not rule directly over the whole universe, the day will come in the future when he will rule directly from the earth over the heavenly bodies through the more continuous development of science. (P.69)

1.4. Summary

From this book, encyclopedias will have to change their definition of the cosmos.

1) Cosmos

Ge 1 : 1 "In the beginning, God created the heavens and the earth"

– Spirit world (Heaven) – Physical world (earth)

a) Spirit world : heavens

Kingdom of Heaven – Paradise – Hell

b) physical world
Ge 1 : 2 A shapeless and empty land

Universe

2) Universe

Space, dark matter, dark energy, Galaxies

3) Galaxies

Stars – Planets – Exoplanets – Other celestial objects

4) Planets and exoplanets

– continents, – waters – gas

The visible Universe contains in its empty space, the galaxies. Galaxies contain stars, planets, exoplanets and other bodies roaming in the vacuum such as asteroids, comets and meteorites. Planets contain continents and water in various forms such as planet earth. The continents of the earth contain the countries. The Divine Principle of the Moon couple describes the universe as the visible aspect of the invisible God.

Chapter IV :
The Big Bang

(2nd Biblical Day -1st era for science)

Engraving demonstrating the Big Bang or the separation of the waters from the waters

God creating Heaven and Earth, 19th century colourised engraving, after Raphael (1483-1520). https://www.la-croix.com/Religion/Spiritualite/Les-*propos-du-pape-sur-le-big-bang-sont-ils-revolutionnaires ?*

In this engraving, God was invisible, so no one could see Him at work. Even if a fictitious witness was present, he would only have seen the manifestations. Hence Stephen Hawking's statement that "the Universe arose spontaneously and that only physics and its laws are at the origin of this phenomenon". (Cf : Stephen Hawking, The Grand Design, published in 2010).

The Universal Cosmological Model, on the other hand, does not state that God created the universe, but attempts to explain how these phenomena occurred, recognising.

The Big Bang being a transitory stage in the history of the universe, one cannot hope to solve all the problems by its theory, nor to end up with the Theory of Everything by starting the analyzes at this level.

For Moses, initially *"The earth was unseen and empty, there was darkness over the face of the deep, and the spirit of God moved above the waters"* (Ge 1 : 2). (Septuagint Version)

Speaking of the earth, he is already talking about an established Universe, despite being empty but with energies, according to what physicists have found, namely the vacuum contains energy. Moreover, it is the presence of the energies of the void that made it different from the spiritual world which is invisible. In the verse : "In the beginning God created the heavens and the earth" (Ge 1 : 1) ; the heavens is the spiritual part of the universe and the earth is the physical part.

We have already explained the Big bang in the first point, saying that it was the sound that was made when God separated the waters from the waters (Ge 1 : 6) and which scholars have called the explosion of the bubble that gave birth to the Universe. Today scientists are starting to talk a lot more about the expansion than about the explosion or the noises. Either way the Bible has used the words "separation" and "expanse." The language is the same except that the Bible speaks of the final stages whereas, science is restoring the scientific names by going step by step, names created several thousand years after Moses, the scholar of the scholars.

Here is a science-based illustration of what has happened from the Big Bang to the present day :

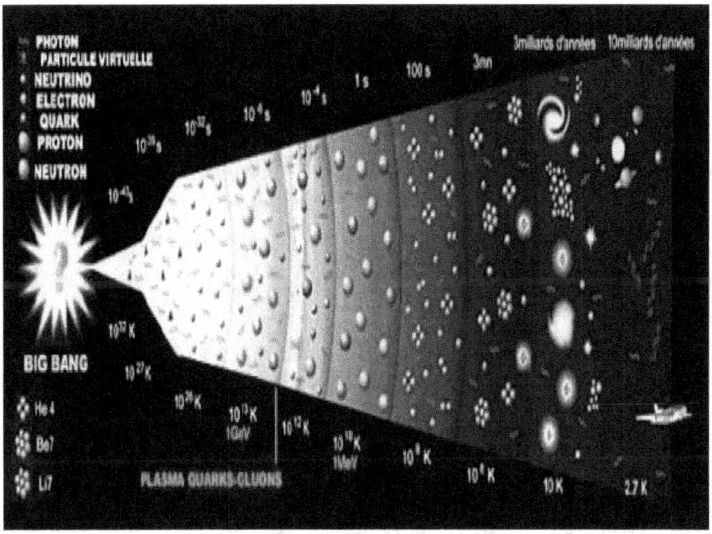

astropolis.fr/articles/histoire-de-univers/l'instant-zero-ou-singularite-initial

The Big Bang Theory is called "standard" by astrophysicists and held up as dogma in the scientific world today. To this day, we too in this book, we monopolize it, but, who has demonstrated to scientists that it was not a coincidence, unless Moses has in his day a laboratory more advanced than anything we have today. Otherwise, let us accept that there is a superior being who witnessed the formation of the Universe and inspired Moses : God, the Creator.

Moses who was his faithful pupil and servant deserves the title of scholar of scholars, because God cannot reveal such material to just anyone and the latter classifies it in a precise order.

To verify if the Big-bang, called the "separation of water by water" in the Bible, really took place, scientists carried out calculations on the stability of the cosmic diffuse background (fossil radiation) and on the expansion of the Universe, observed by the remoteness of the galaxies by the astronomer Hubble. If there is expansion; there was a point of departure, concluded the Belgian scholar Abbe Georges-Lemaitre.

Their conclusions were that the Big Bang actually did take place, but this book is the first to relate it to the history of biblical creation.

No one had ever thought that the knowledge of the Big Bang was already ancient and contained in the Bible, or even later in the writings of the Greek Philosopher whom I have just resuscitated in science and whose name is Empedocles.

Section 1 : The Big bang or "the separation of water from waters"

Ge 1 : 2 The earth was formless and empty : there was darkness over the face of the deep, and the spirit of God moved over the waters.

Ge 1 : 6-8 "God said, let there be an expanse between the waters, and let it divide the waters from the waters.7 And God made the expanse, and he divided the waters which were in the waters. below the expanse from the waters which are above the expanse. And so it was.8 God called the expanse of heaven. So there was evening, and there was morning : it was the second day".

1.1. The origin of the word BIG-BANG

As said before, previously the history of the universe was a sacred material and no one could say anything contrary to the biblical creation story without the Church getting involved. It was first the Arabs who began to introduce the scientific reasoning contrary to the Bible before the currents reached Europe.

Indeed, having conquered the Crusader Wars, Muslim women no longer felt bound to papal authority, hence they could clearly state what he thought scientifically. It is not the purpose of this book to talk about it in detail, and we get to the point.

The first cosmological theories appear in Mesopotamia around 4000 BC. J.C. According to these theories, closely related to the religious beliefs of the time, the Earth is the center of the Universe and other stars move around it. This theory is repeated by Aristotle and by the Greek astronomer Ptolemy who maintained that the movement of stars in the sky results from their fixation on spheres rotating around the Earth.

Thus, at the very beginning of the 2nd century, Ptolemy (c. 100-170), astronomer, mathematician and geographer of Greek origin, whose theories in astronomy dominated scientific thought until the 16th century confirmed the idea that earth had a central position in the Universe and that the other planets revolved around it. His position was never contested, for it fitted, according to the church, to the story of Creation where the formation of the earth preceded that of the stars (Ge 1 : 9-16) ; That is, all the stars were aligned to support life on earth.

The revolt within the Catholic Church, which gained momentum in the 16th century, allowed some scholars of revolutionary nations to publish their theory. Thus, based on question marks emitted by Ptolemy himself, to whom he offered unconvincing answers, which we could not carry over into this book so as not to turn it into a manual of debate, the Polish astronomer Nicolas Copernicus (1473-1543) rejected this theory, and developed his own theory, called heliocentric, which affirmed on the contrary that the Sun was located at the center of the Solar System and that all the planets, including the earth, revolve around it.

Copernicus's system was immediately the subject of much criticism : it had only a dozen defenders, the most famous of which were the Italian scholar Galileo and the German astronomer Johannes Kepler. Despite the invention of the astronomical telescope in 1609 by Galileo which confirmed the Copernican system, the Church opposed this theory of a mobile and non- central Earth, which supposedly contradicts the biblical diagram on the origin of the world. In 1616, Copernicus's work was banned. Galileo was accused of heresy and was tried for it by an Inquisition tribunal in 1633 where he was condemned. In October 1992, a papal commission recognized the error of the Vatican and rehabilitated Galileo.

In addition to other key discoveries he made, Galileo remains a symbol of the battle against the authorities to free science from the influence of religion. (Cf : Galileo : Microsoft® Encarta ® 2009. © 1993-2008 Microsoft Corporation).

In the twentieth century, freedom of thought being widespread and protected, people began to argue not only about the functioning of the universe, but also about its origin without sticking to the biblical schema which will be rejected besides and downgrade. This time it was no longer just a question of philosophical or religious thoughts, but of science. No book on astronomy or cosmology makes more reference to the Bible, even historically in terms of the origins of the Universe. It is my publications that put God back on the scientific scene. And this is the reason that made God interested in this work.

Albert Einstein's discovery of general relativity in 1915 marked the beginning of modern or classical cosmology, where it becomes possible to describe the Universe as a whole as a physical theory.

Of all the ideas that arose in the 20th century about the origins of the universe, two fought for place for half a century. It is that of the stationary universe, defended by the British scientist

Fred Hoyle and of the primitive atom of the Belgian scientist, Abbe Lemaitre.

To confront the two schools, that of the "cosmological principle" defended by Lemaitre and that of the "perfect cosmological principle" defended by Fred Hoyle, the two protagonists found themselves on the set of the BBC in the 1950s; a Radio channel that is no longer to present. It was during the program, in total disagreement with the theory according to which the Universe was born from an explosion from an original atom, that the English astrophysicist Fred Hoyle uses the term "Big- bang" in the tone of mockery, to designate with irony the notion of *"primitive explosion"* introduced into the cosmological model, developed in the 1920s by the Belgian astrophysicist Georges Lemaitre and the Russian physicist Alexandre Friedman.

This colorful expression, used to make fun of those who said that the universe started with a bubble that expanded, will then become embedded in scientific vocabulary. And it was George Gamow who first set up the physical theory of the Big Bang.

The big bang theory was introduced in 1948 by Russian-born American physicist George Gamow, modifying Lemaitre's theory of the primitive atom. Gamow assumes that the Universe was born from a gigantic explosion, and that the various elements observed today were generated just after this explosion called the big bang. It is this theory that is recognized today as the Standard Model.

In addition, it is this idea of explosion and not of expansion, which has led to the discovery of many things which today constitute the scientific proofs of the Big-bang or of the "separation of water from waters", such as the residue of the first light and gravitational waves.

1.2. Pope Francis and the Big-bang

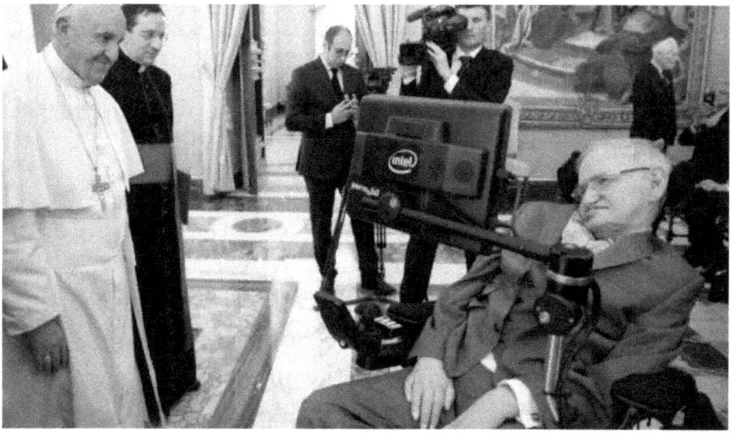

The atheist physicist Stephen Hawking and Pope Francis or the meeting between religion and science : "we must demolish the walls and build the bridge" (Pope Francis). Astrotheology is an answer to this call.

We cannot pass under silence, the words of the Pope Francis who witnessed the publication of this research.

On October 27, 2014, Pope Francis made the following remarks before the Pontifical Academy of Sciences "*The big bang does not contradict the creative intervention of God, on the contrary, it requires it.*" *(Cf : https://www.la-croix.com/Religion/Spiritualite/Les-propos-du-pape-sur-le-big-bang-sont-ils-revolutionnaires ?)*

This intervention of the Pope opens a new era on the schema of the universe for scientists as well as for religious people. These words have provoked many reactions simply because scientists brandish the Big Bang to dismiss God as Creator. But with the Universal Cosmological Model, we discover that this word created in 1950 is indeed biblical, and known as the "Separation of the waters from the waters" (Gen 1:6).

This scheme from the primitive atom to the Big Bang was already endorsed by Pope Pius XII in his address to the Pontifical Academy on 22 November 1951. The theme of his speech was "the proofs of the existence of God in the light of the present science of nature". As we read earlier, Abbot George did not appreciate his speech.

The problem was that no one discovered the verse that demonstrated that the universe had experienced a Big Bang during its evolution. So the discovery of this verse was for me the greatest discovery, because it took place at the second stage of the universe and not at its beginning. This really makes the existence of the pre- big bang which is described on the first day in the Bible. We explained this sphere at length in the first part of Volume I.

This biblical discovery that the Big Bang took place at the second stage of the Universe breaks Stephen Hawking's statement that "asking what was before the Big Bang was absurd since it was like asking what is south of the South Pole. This biblical discovery rather confirms the work of scientists who have described the Big Bang as a rebound such as the German physicist Martin Bojowald and those who have described the pre-Big Bang as a moving sphere such as the Italian physicist Gabriele Veneziano.

These biblical discoveries break several statements made by Stephen Hawking. Stephen Hawking had declared : "the discovery of the expansion of the universe" - the Big Bang theory - was one of the most important intellectual discoveries of modern times. *(Cf : https://fr.Aleteia.org USA - Published on 14/03/18).*

I would add that the location of this discovery in the Bible was also one of the most important in theology, because it now unites science and religion with regard to the scheme of the creation of the universe. The other facts, discovered in this scheme in the Universal Cosmological

Model, will be validated later.

In his speech of 27 October 2014, Pope Francis made other very important statements which are confirmed in my scientific arguments in the books of the collection ”Verses of the Scholars”, although he had no knowledge of them, and I too had no knowledge of these statements. Among these statements we have such statements as :

– *The Big Bang was an act of love from God. Science and its discoveries do not question the existence of God : on the contrary, evolution (for example) is part of God’s design for His creation.*

– *The beginning of the world is not a result of a chaos that owes its origins to another chaos, but derives directly from a supreme Principle that creates out of love.*

– *The theory of evolution in nature does not contradict the notion of Creation, because evolution presupposes the creation of beings that evolve.*

Section 2 :
The Real Evidence of the Big Bang

2.1. Scientific evidence of the Big Bang

Currently; the big bang theory provides a satisfactory framework for the evolution of the universe. Today we speak of the three observational pillars that support the big bang or the separation of waters by waters (Ge 1 : 6) according to the Universal Cosmological Model.

First, there is the radiation from the cosmic diffuse background, considered to be the first remnants of light that arose after the Big Bang. Predicted in 1948 by the American physicist of Russian origin George Gamow and detected in 1965 by the American astrophysicists Arno Penzias and Robert Wilson, the cosmic diffuse background is interpreted as a trace and proof that the primordial Universe experienced a very dense phase and very hot. The Bible also confirms this by speaking of the appearance of light before the Big Bang in Ge 1 : 3 (... And! And there was light), which shows that there was a warming of the original elements. We talked about it. We talked about this in Volume II, which is entitled "Before the Big Bang to Fossil Radiation".

According to the big bang theory, this microwave radiation at 3 K is a residue of the energy prevailing at the time of the formation of the first stable atoms in the Universe (decoupling material radiation) - or about 380,000 years after the primary explosion. This fact of observation today constitutes one of the pillars of the big bang theory, the most likely theoretical model for describing the evolution of the Universe.

Second, it is the abundance of heavy elements that is

forming from light elements starting with hydrogen to helium and so on. This shows that there has been a unique and evolutionary origin for the formation of materials in the universe.

Third, it is the expansion of the Universe detected by telescopes. And, the astrophysicists said to themselves, since the universe flourishes, there has been a starting point. Hence the Belgian scholar, Abbe G. Lemaitre had put forward the idea of the primitive atom which exploded, we have said.

The Universal Cosmological Model has just added a fourth pillar understandable by all as proof that there was indeed the Big Bang. This is the man. Indeed, astrophysicists say that man is the dust of the stars, while the Bible supplements by saying that man is the dust of the earth. This is because the earth, according to scientists, was formed by protoplanetary dust from a star. And, this dust according to the bible, will result in the formation of man by God. The proof is that all the chemical elements found in humans are found in the earth and in the stars. In short, everything came from a single source element which was the primitive atom underwent an expansion called the Big-bang and was the basis of the constitution of all the elements and structure of the universe.

By that, we end up with the veracity of the biblical verses which say that, at the beginning there was an original material described as waters (Pre Big-bang) and these waters were separated (Big- bang). With the waters below, the earth will sprout, out of which will be formed living beings, and with the waters below, God was the thresholds. (Ge 1 : 2-9).

2.2. The confirmation of the Big-bang by the bible

Is science confirming the Bible or is it the other way around? In principle, science should confirm the Bible if the goal of scientists was to verify the biblical creation story. But since the biblical story had no scientific basis, no scholar was interested in it. Not even that, without the knowledge of the "Principle of creation", branch of the divine Principle of Rev. Dr. Sun Myung Moon, the creation story of the Bible would be buried by science and relegated to antiquity, for no one, even believing astrophysicists, was ready to wave it before scholars.

I have watched a lot of documentaries from those who defended God in the face of scholars, but no one was up to it and had compelling arguments. All appealed only to faith and had only one question, to which they themselves could not answer, "what had he before the Big Bang", to justify divine intervention.

Thus the Universal Cosmological Model, which is the scientific restitution of the history of creation, has just demonstrated that the Big-bang is biblical. This raises the question of whether it is God who is the Creator of the universe. In a way, the Bible has just confirmed the first stage that scholars have reached, by confirming the Big Bang. All Christians have read or waited for the above verses before, but no one for a century has had to relate to the Big Bang that scholars used to wield to dismiss God as Creator.

The second step was the discovery of the cosmic diffuse background which is the first light that appeared before the Big Bang, confirmed by its current temperature which is uniform throughout the universe.

As we can see, the notion of the origins of the Universe has become a big issue for scientists. The clerics and their opinions had no place without a scientific demonstration based on observation. As we have said, all research on the Universe leads to God. But, to avoid meeting God and confirming his existence, scholars stopped even before the Big Bang. Read their arguments :

"Since no theory describes the universe below Planck's time, the Big Bang itself remains inaccessible to scientists. The very origin of the Universe therefore remains an indecipherable enigma. Moreover, the Big Bang theory does not describe the events immediately preceding the primordial explosion. Time itself may begin at the instant of the Big Bang; attempting to describe a "Pre Big-bang" would therefore make no sense. " (1) (1) Big-bang theory : Microsoft® Encarta ® 2009.

For scholars, what we are developing here made no sense. Fortunately for me, the Bible was written before their conclusion, 1500 BC. JC. Hence for the moment they are forced to reverse their deduction concerning the Pre Big Bang. Thus, by discovering the scientific proofs of the Big Bang, the scholars come to what the Bible says in the diagram of creation as we can read them in the following verses :

"God said, let there be an expanse between the waters, and let it divide the waters from the waters.7 And God made the expanse, and divided the waters which were under the expanse. from the waters which are above the firmament. And it was so". (Ge 1 : 6-7)

2.2.1 The separation of the waters from the waters and the Big-bang

Moses used the word "separation" in place of the original mass shattering to demonstrate that the phenomena dubbed "Big- bang" by science was under divine control and that it was no accident. When you purposely crush something in order to collect what is inside, you cannot speak of an accident or a Big Bang, however long it takes. But, if you step on it unintentionally, even if you collect what is inside, it constitutes an accident, it is the Big Bang. (Gen 1:6)

Let's read how Moses or the Bible talks about Big-bang :

Ge 1 : 2 The earth was formless and empty : there was darkness on the face of the deep, and the spirit of God moved over the waters ...

Gen 1 : 6 God said, let there be an expanse between the waters, and let it divide the waters from the waters. 7 And God made the firmament, and he divided the waters which were under the firmament from the waters which were above the firmament. And so it was. 8 God called the expanse Heaven. So there was evening, and there was morning : it was the second day".

We take as key words in all these verses "the separation" and "the expanse" that is, the empty space created by the separation of a bubble called water by Moses. Today some scholars prefer the term "expansion". This phenomenon of the origin of the Universe is described by scientists as the Big Bang. In the beginning we have said, this word was to make fun of those who supported the theory of a dynamic Universe and in the end it was to push God aside. Fred Hoyle was the first to pronounce it and the Russians Friedman and Gamow in theory passed it, we have said.

Moses as we have said, used the term "God separated" to demonstrate that it was no accident of nature, even though it was a physic-chemical consequence. But science uses the term Big- bang as if to say that it was a compression of the universe which burst apart exploding under several debris forming galaxies without the intervention of any intelligence.

The same scientists, after having observed in those days

that there is a harmonious hierarchy of the stars, still want to question the Big-bang theory, because, they say, an accidental bursting cannot create such an order. We tell them that this is how they must see the hand of God whose presence Moses originally mentioned when he said that in the beginning God created the heavens and the earth.

Take for example a simple case. In the quarries of stones or mines, to clear a path or to exploit it, explosives are used. For an ordinary man who is not in the secret, this is just an explosion of fires and can spread the word. But for whoever was in the secret, he will say the operation was carried out well, focusing on the path and not on the debris.

The Big Bang created a lot of debris, but for God it was the formation of the Milky Way in general and the Earth in particular that was targeted. This is why, Moses who knew the goal that God was pursuing, speaks of the "separation of the waters from the waters", while the scholars, who are far from the secret of God, speak of the Big Bang. We tell scientists that all phenomena from before the Big Bang to the post Big Bang had our existence as its end. Thus, after having created man last, God exclaimed "this is very good" and, granted himself a rest on the seventh day (not to be confused with our 24-hour days as explained in 'introduction).

Gen 2 : 3 God blessed the seventh day, and he sanctified it, because on that day he rested from all his work which he had created while doing it.

Going back to our example of the explosion of a quarry or a mine, we cannot think of a well-sculpted form of the debris, because each (galaxy) will have its dimension. But for the blaster, it's the goal that counts. He will pay more attention to what he is looking for. He is going to carve it according to the idea it has fixed in his heart. This is why after the Big Bang, God directly took care of what would become of the earth before taking care of the star and the planet which will be of primary importance to it. Read :

Gen 1 : 6 God said, let there be an expanse between the waters, and let it divide the waters from the waters. 7 And God made the firmament, and he divided the waters which were under the firmament from the waters which were above the firmament. And so it was. 8 God called the expanse Heaven. 9 God said, Let the waters which are below heaven be gathered together in one place, and let it be dry. And it was so.10 God called the dry land earth, and he called the cluster of waters seas. God saw it was good.

Gn 1 : 14. God said, let there be lights in the expanse of heaven, to separate the day from the night; let them be signs to mark the eras, days and years; 15 and let them be lights in the expanse of heaven to give light on the earth. And so it was. 16 God made the two great lights, the greater light to rule the day, and the smaller light to rule the night; he also made the stars. 17 God set them in the expanse of heaven to give light on the earth, 18 to rule over the day and over the night, and to separate the light from the darkness. God saw it was good. 19 So there was evening and there was morning, the fourth day.

As we can see, Moses was concerned with speaking only of what is essential for the teaching.

The Universal Cosmological Model accepts the scientific name of the Big-bang phenomenon, but since there was a precise goal, namely to achieve the creation of an environment conducive to living beings, we reject the idea of an accidental explosion to say that there was the intervention of an intelligence - God. The same explosion or damage may or may not be accidental. It is accidental if it is unintentional, but, if that was the goal, it is not accidental. This is how Moses used the phrase "separating the waters" rather than exploding the waters. Warning! Speaking of God, I also do not exclude what really happened on the physical plane, that is to say the contribution of the energetic forces that we will be analyzed in the previous volumes.

Nothing can now separate the biblical viewpoint on the beginning of the visible universe with science. There was indeed an original "water" bubble and then came the "Big-bang" separation. But, it was the Bible that was the first book to state it, not science. And yet, no Biblical reference can be found in scientific journals dealing with astronomy and cosmology. From now on, it will not be so.

The Universal Cosmological Model has just made religion and science work together with regard to the origin of the Universe, and direct scientists to continue the work of the seventh day of creation according to divine recommendation :

Ge 1:26 Then God said : let us make Humankind in our image, in our likeness, and let him rule ... over all the earth (the universe)

2.2.2. The end of the mythology of the creation scheme biblical

To this day, cosmologists describe the creation story of Genesis as metaphysical. But, as I have just demonstrated, Moses actually demonstrated the pattern of the origin of the universe. As a reward, it is called a myth, as the great Greek deities continue to be immortalized with the names of the planets. We don't even pay attention to Moses, who described the order in which the Universe was created, from before the Big Bang to Humankind. His sin for science was to have recorded this knowledge in a religious book and to have mentioned the name of God.

It should be noted that in the attribution of the names of the planets, the Greek mythology is largely glorified by the

International Astronomical Union (IAU). Let us take only the case of the planets of the solar system :

– Mercury : messenger of the gods

– Venus : goddess of beauty and love

– Mars : god of war

– Jupiter : Roman god ; master of heaven and earth

– Saturn : Father of Jupiter, god of sowing and of the vine

– Neptune : god of the sea

– Pluto : Greek god of the underworld

As for the earth itself, the name comes from the god Aidônéus (Ha-dés), god of the underworld. Indeed, Lucifer tore him from the false parents Adam and Eve and he arrogantly denied him to Jesus (Lk 4 : 5-6), and who did not contest him. Jesus Christ only reminded him of the law (Lk 4 : 8). Read :

Luke 4 : 5-6 The devil lifted him up, and in a moment shewed unto him all the kingdoms of the earth, and said unto him, I will give you all this power, and the glory of these kingdoms ; for it has been given to me, and I give it to whomever I want.

Luke 4 : Jesus answered him, it is written, You shall worship the Lord your God, and serve him alone.

Even the prophet Elijah who was the first to create a propulsion motor as we will see in the other volumes, was never re-compensated by the name of a probe or a rocket, because his story is contained in a sacred book and not in mythologies.

Generally speaking, it is the International Astronomical Union (IAU) that has the right to officially name celestial objects. So far we've seen this :

– The planets of the solar system bear the names of the deity of the Greek or Roman pantheon ;

– Exoplanets are named after the deity of some pantheon ;

– Comets are generally named after the names of their discoverers. In the story of Genesis, chapters one and two, Moses gives

God the place that is due to him, that is to say the only actor in creation. But starting from the exodus which will be the subject of the other volumes, we will see Moses himself at work by confirming his unparalleled supremacy of scholars since he had the idea of bringing out the children of Israel in Egypt and make it a first heavenly nation.

The Big Bang alone occupied a whole day according to the Bible and a moment according to science. Even for the Bible, it only occupied a moment, but since it was a phenomenon of great importance which marked the end of a cycle of the universe and the beginning of a new cycle, and, wanted to give this event greater importance, Moses called it a day.

Ge 1 : 6 God said, Let there be an expanse between the waters, and let it divide the waters from the waters. 7 And God made the firmament, and he divided the waters which were under the firmament from the waters which were above the firmament. And so it was. 8 God called the expanse Heaven. So there was evening, and there was morning : it was the second day

Looking at the size of the Universe, if all galaxies were brought together with at least 200 trillion stars, one wonders how God would have formed such a compact cluster before the Big Bang. All scholars now know that this was not so in the beginning. Some scientists even assume that the universe was only in the form of a small atom which ; after explosion or expansion, by gravity, each particle began to attract other particles that were scattered in space. By combining each in its own sphere, and this is how, passing very hot stars described as the first stars, other galaxies were formed afterwards.

Thus, we declare the Big-bang or separation of the waters from the biblical waters as the metric Unit of the history of the Universe both for science and in the One-verse Cosmological Model, and baptize it under the name of ”the zero moment of the Universe”, as it is with the birth of Jesus in Christian history. As I wrote this statement, I was unaware that science had also come to the same conclusion and the same terms.

Chapter V :
After the Big Bang

(3rd and 4th biblical days)

The post-Big-bang represents the period which followed the zero moment of the Big-bang or the "separation of the waters by the waters" according to biblical terms. According to the Principle of creation, taken up in the physics of creation, the palpable materials which followed this state began with minerals on the earth, plants, animals and men. It fits with the biblical pattern of creation. Some, except astrophysicists, will tell me that the Bible talks about water.

Indeed, water is hydrogen and oxygen, so we are dealing with minerals. And these are the minerals that began to form the first stars.

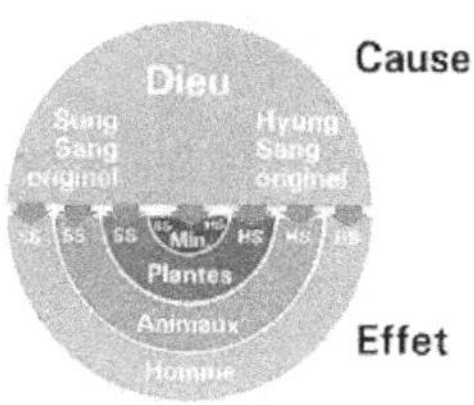

Source : Divine principle level 4, p18

Section 1 :
The formation of stars according to Universal Cosmological Model

According to the Microsoft Encarta 2009 dictionary, one expects by star any natural celestial body, whether or not luminous, representing a physical reality.

According to science, the Big Bang was followed by a period when the heat began to lose its intensity and the slimy material began to compact. This was followed according to science, looking in the comparative table that we appended at the beginning, by a period of the formation of atoms of hydrogen, helium, and later other single atoms, cold clouds of gas cosmic, including galaxies that the Bible already calls the formation of the earth, for what other name could Moses use to make himself understood by the audience of his day?

1.1. Design of the word "earth" before and after the Big Bang

We have said that before the Big Bang the word "earth" was used in the Bible to refer to the universe and that through the verse "the earth was formless and empty" (Ge 1 : 2). And we said that something that has no form and is empty is like nothingness, and was none other than the universe when it started. It was in the void that many phenomena that we are taught in quantum mechanics that are part of the physics of creation unfolded.

For the Universal Cosmological Model, the Universe did have a beginning, but initially there was a vacuum until cosmic deflation had to condense all the energy into one point. Then there was the Big Bang, followed by cosmic inflation.

Because of cosmic inflation and the extreme heat that reigned just after the Big Bang, the various laws and theories of quantum mechanics cannot be applied there, according to the Universal Cosmological Model. For the Universal Cosmological Model, these different laws and theories can only find their application in an empty universe therefore, even if their authors or creators do not agree, the laws and theories of quantum mechanics concern the Pre Big-bang that is to say the moment when the Bible spoke of a shapeless and empty earth (Ge 1 : 2).

Indeed, at the beginning, cosmic inflation was characterized by a repulsive force of the particles, which resulted in the expansion of the universe, before the latter cools and the Quarks assemble to form the first elements, according to the current scientific conception.

This is why the laws and theories of quantum mechanics do not match the theory of general relativity which explains the universe after the Big Bang.

Currently, thanks to quantum physics, we are able to create a lot of things, because it is close to the physics of creation; part of the Principle of Creation of Rev. Dr. Sun Myung Moon dealing with the universe.

It should also be noted that I am the author of the name of the Physics of creation, because astrophysicists equate the word "Principle" with a philosophical or religious notion. Concretely, all modern electronics that we use today are the result of research in quantum mechanics. From the invention of the transistor and its evolution to the microprocessors which are now an integral part of our life.

Coming back to our topic, according to quantum mechanics, in a vacuum there are particles and we stick with that. This is also what the Principle of Creation confirms by speaking of the Universal Prime Energy which has passed from spiritual to physical, constituting in the process a material sphere described as formless and empty earth. This is where all the dances we are taught in quantum mechanics have taken place.

In fact, at the moment, we don't even know whether to use the empty terms, chaos, nothing or none. In astrophysics all these terms no longer have their literal meaning in the composition of the universe.

In the post-Big-bang according to the Bible, we are still told of the earth, not as formless and empty (Ge 1 : 2), but as a dry mass "God called the dry earth, and he called the heap of the waters of the seas" (Ge 1 : 10). Here we see that the material has given way to the immaterial and we are in the post-Big-bang period. (Gn 1:6- 9). Read :

Ge 1 : 2 The earth was formless and empty : there was darkness over the face of the deep, and the spirit of God moved over the waters. (Before Big-bang).

Ge 1 : 6 God said, Let there be an expanse between the waters, and let it divide the waters from the waters. 7And God made the firmament, and he divided the waters which were under the firmament from the waters which were above the firmament (the Big bang).

Ge 1 : 9 God said, Let the waters which are below heaven be gathered together in one place, and what ! the dry appears. And so it was. (After Big-bang) .10 God called the dry land earth, and he called the cluster of waters seas. God saw it was good.

The keywords here are the waters in one block, the separation (expanse and expansion) of the waters into two blocks, the formation of the earth from the waters which lie below the expanse.

Why, after the Big Bang, did Moses speak only of

planet earth? Because he's the only one that could be a source of interest to his contemporaries. The other galaxies presented themselves as stars and were referred to as luminaries (Ge 1:14). In fact, it was not until the 1920s, thanks to Edwin HUBBLE, that science became aware of other galaxies other than the Milky Way and this appears as luminaries to the naked eye.

Right after the Big Bang or the separation of the waters from the waters, God directly concerned himself with a flower bed point that was dearest to him; the earth. This was the planet that he decided to materialize. To that, other planets and galaxies will have to be positioned in such a way as to bring life to this one. This is also scientifically taught in the Anthropic Principle. This has prompted some thinkers to say that the earth is at the center of the Universe. But, as for us, we say that the earth is at the center of the heart of God. To say that the earth is at the center of the Universe was just a bad statement of that sentence.

1.1.1. The origin of the earth

Speaking directly of our planet earth after the Big Bang, the description that Moses was for his training can only surprise us for someone of his time, that is to say 1,500 years BC. Read :

Ge 1 : 9 *God said, Let the waters which are below heaven be gathered together in one place, and let it be dry. And it was so. 10 God called the dry land Earth, and he called the cluster of the waters Sea. God saw that it was good.*

According to the Big Bang, during the period of galactic collisions, things were sometimes in the form of magma where we found the mineral, non-mineral and gaseous elements as we can read in this article from FuturaSciences.fr :

"Indeed, at the very beginning of the Hadean, which began about 4.56 billion years ago, our Planet was formed thanks to collisions and captures of celestial bodies, some of which exceeded a hundred kilometers from diameter. Becoming more and more massive, its attraction increased, giving rise to increasingly violent shocks, releasing heat capable of melting the rocks. Moreover, it is believed that the bombardment converted so much gravitational potential energy into thermal energy, that an ocean of magma several hundred kilometers deep, and perhaps a thousand, existed on the surface of the Earth ". (Cf : Le 07/08/2016-Laurent Sacco, Futura-Sciences).

In other scientific explanations, dry land emerged as the formation of islands now. That is, the dust having accumulated as a single lump in the water, condensed to form the dry part that we now call dry land. The difference between religious and scholars is that for the latter it was processes due to chance and obedient to certain physicochemical processes. But, for the religious, despite this obedience to these processes, it was in the plan of intelligence with a specific purpose as mentioned in the verses above.

In fact, in the Bible, we are told that there were first the waters and that in these waters there was the process of the formation of some dry things which will become the earth, due in particular by the agglomeration of dust. That's a way of saying about the beginning and the end, but the process was more complex than one can imagine.

In the scientific documentary "Keys to the Universe 4 - The Origin of Water on Earth" we are told that at the very beginning of the formation of our solar system, there was a vast cloud of gases and dust that floated in space, but inside there were large amounts of hydrogen and oxygen. After thousands of years under the action of cosmic rays, these atoms combine to form water (H2O). Inside the gas cloud, water molecules stick to grains of dust, then freeze, forming screeches of ice. The cloud of frozen water becomes so dense that it eventually collapses under the effect of its own gravity; this is the start of the process that will spawn our entire solar system. This is the scientific scheme for the formation of all that is dry and gaseous in the solar system including our planet earth.

We see that this diagram indeed fits with the scientific restitution of the biblical verses above (Ge 1 : 9-10).

Referring to the time of Moses, how could he conceive of a logical creation story? It was only the Creator, the designer of these things that could inspire him, and that designer is none other than God : our Father. And, he can dictate these things according to the principles of predestination (see chapter 6 of the divine principle) only to a scholar who can understand it, whether he is religious or not.

1.1.2. Cosmological parallelism between the bible and science

ACCORDING TO THE BIBLE			ACCORDING TO SCIENCE (COMPILATIONS)		
CYCLE	DAY	CREATION	**CYCLE**	ERA	APPEARANCE
Before Big- bang Ge 1 :2	1st	— — darkness (dark energy — abyss (black hole) — Spirit of God (energies) — waters (original particles) light (lights, heat, rays))	Before the Big- bang	- aporia	Gray area for science. No scientist dares to express a point of view for fear of being ridiculed one day. But, several assumptions are starting to hold true with Ge 1: 2 including string theory and others that imagine the Big-Bang Before. .
Big- bang Ge 1 :6	2e	- Separation of the waters from the waters	1st	- 15,000,000,000 years	Big bang (explosion or expansion of the universe) followed by the formation of the elements.
After Big- bang Ge 1 :9	3rd	Material Universe: dust, planetesimals, gas,life molecules	2nd	- 14,999,700,000 years at - 14,900,000,000 years	Formations of helium atoms, hydrogen atoms, cold clouds of cosmic gas, galaxies

Ge 1 :14	4th	formation of stars and planets	3rd	- 10,000,000,000 years	formation of stars and planets
Ge 2 :5 Ge 1 :20	5th	First plants Sea and flying animals	4th - paleozoic	- 4,100,000,000 years to -400,000,000 years	— 1st chain of nucleotides ,- shallow oceans, carbon -based compounds, — first bacterial cells, DNA, — complete development of bacteria, the oxygen level in the atmosphere reaches its current level: 21% — first plants (Paleo-zoic) —First land animals (insects, amphibians)
Ge 1 :24	6th	Mammals and humans	5th Mesozoic-Jurassic	- 200,000,000 years to - 4,000,000:	- flying reptiles, 1st mammals (Mesozoic) - flowering plants, first primates (Juras-sic) - 1st man (Australopithecus)

Sources : compilations

Section 2 :
The molecules of life

2.1. Chronological contrast between the 3rd biblical era and the 2nd scientific era

Many people who read the Bible very carefully, especially atheists, make this point to say that the biblical pattern of creation is not real, for how plant life may have appeared before the lights?

Indeed, in the table above, there is a discrepancy between the 3rd day of creation and the 2nd cycle of science regarding the formation of the earth and the appearance of plants before photosynthesis, but, this is a misunderstanding.

Let's read this verse first :

Ge 1 : 12-13 "The earth brought forth greenery, and grass bearing seed after its kind, and trees bearing fruit, and having their seed within them after their kind. God saw it was good. So there was evening, and there was morning : it was the third day ".

We have said that Moses spoke of the initial stage and the final stage (then there was evening and there was morning). Due to the fact that after the Big-bang, the galaxies began to form (as science attests in the second stage), the earth, being a galactic element, was already included according to God's purpose.

Describing the germination of plants at this stage referred to the formation of molecules of life at that early stage in the history of the universe. This means that the same molecules can also be found in other stars, but while being static, while those of the planet earth will bear the fruits. We will come back to this. (See the appearance of organic molecules).

Thus today, the existence of organic elements in certain stars begins to complicate certain theories of exobiologists, in particular those elements which are today the basis of the composition of living beings. As we have said, Moses had already designated these elements on the third day of creation by molecules which he described as plants in Ge 1 : 11-12. Even in the Ancient Eastern Models - which we will talk about later, these elements were referred to as wood. But, the plant kingdom, just like all living kingdoms, began to appear from the fifth day of creation as described in Genesis 2 : 5.

Ge 2 : 5 "When the LORD God made the earth and the heavens, no shrub of the field was yet on the earth, and no grass of the field was yet sprouting : because Eternal God had not made it rain on the earth, and there was no man to cultivate the soil".

We will come back to this when we talk about the living kingdom, as well as the appearance of the molecules of life in the chapter dealing with the post-Big Bang. In addition, with the Universal Cosmological Model, science will now have to rectify its cosmological cycle by taking into account the period preceding the Big-bang where very important activities had taken place.

2.2. The appearance of molecules of life

In the scientific documentary "The origins of the earth"; we are told that at the origins of the earth, the earth's crust was already the bearer of life while they were still undergoing all kinds of bombardment. Even before the moon separated from the earth, according to the same documentary, there were already the molecules of life. Hence, it is the projection that Moses was speaking of the earth. It is a documentary that I recommend to scientists who also evoke as a mystery the origin of the water to which we will return.

In fact, in the Bible, in speaking of the creation of the earth on the third day, we have said, Moses also speaks of the appearance of plants, whereas according to science, plants appeared at a time corresponding to the 5th biblical day. We have said that they were molecules that brought life to this planet, as we can read :

Ge 1:9 " *God said, Let the water under the sky be gathered into one place, and let the dry land appear." And it was so. 10 God called the dry land "earth," and the gathering of the water he called "seas. And God saw that it was good. 11Then God said, Let the earth produce vegetation : seed-bearing plants and fruit trees on the earth bearing fruit with seed in it according to their kinds. And it was so... 13 And there was evening and there was morning : the third day.* "

Indeed, these verses seem to be in contradiction with science as far as the appearance of plants in the 2nd cosmological cycle or 3rd day according to the Bible is concerned. Going to chapter 2 of Genesis; Moses spoke of the appearance of plants on the 5th day, which corresponds with science. Is there a contradiction between these two verses? No, today even the scientists know that in the supernovae that came at the end of the life of the stars, there is the liberation of the atoms that formed inside, including the molecules of life.

This is how Moses, already knowing the finality of what will happen to the viscous mass that will become the earth, speaks of the plants on the third day, which were nothing more than molecules that will evolve to form life, and the latter began with an underwater being, baptised LUCA by biologists, which in its evolution subdivided into bacteria, eukaryotes and archaea. It is the eukaryote, (LECA, Last

Eukaryotic Common Ancestor) which is considered by biologists to be our last common ancestor, starting with an underwater flora, passing through plants, fungi and animals (Cf. : The Challenges of Science : The Origin of Life, edited by RBA coleccionables, S.A.U. 2018, Barcelona, Pp.105-124).

Moses in this verse already unveils God's plan for planet Earth. As for the creation of the plants themselves, this took place on the fifth day as mentioned in the second chapter of Genesis.

I want to come back to this when I talk about the appearance of plants.

What confirms the idea that life-bearing molecules were already present during the period of galactic collisions is the discovery of organic molecules in comets and some planets that were formed or that were ejected at the same time as the Earth. So, those scientists think no more than discovering in some planets or exoplanets outside of the earth organic material, that this will destroy the pattern of creation of biblical genesis.

If in certain planets, comets or stars, these materials have remained static, on earth, there has been evolution until life and this, as we have said, like scientists, by the privileged position occupied by the earth in relation to other stars.

Let's read for this the comments taken from the internet on the organic molecules of comet 67P/Churyumov-Gerasimenko after the analysis of Philae's data by the researchers of the CNRS of France, published in the edition of 31 July 2015 in the journal Science :

"Since the materials of comets have been very little modified since their origins, this means that in the early days of the Solar System, organic compounds were already agglomerated in the form of grains and not just as small

molecules trapped in ice as previously thought. It is such grains that, when introduced into the planetary oceans, could have favored the emergence of life there".

This is a scientific idea that is in line with what we are saying.

Until now, the emergence of this life has only been found on planet earth, even if we are now beginning to speculate on certain stars.

What we should remember in this section is that after the Big Bang, there was also the formation of organic elements in the first groupings of material. This is what makes it possible to find it everywhere and why not activate them since God has assigned man as the lord of creation? For example, why not bring a watering can or a gas-to-water transformer for meteorites? Or not to plant these molecules of life on earth? Cultivating here does not mean sowing, but I am talking about medical culture. The ball is in the court of science.

Section 3 : Lighting

3.1. Galaxies - Stars - Planets (4th day)

By referring to the scientific documentary cited above in *"The keys of the universe 4 - The origin of water on earth"*, when we speak of the formation of the solar system, we see the Sun and the celestial bodies which revolve around it, in particular the planets and their satellites. And in this documentary, we have as a source the hydrogen which will form the Sun as a great luminary on one side and on the other side the hydrogen and the oxygen which by capturing the dust, will form the other bodies. celestial, which appears seen from the earth, like small luminaries by their reflection of the solar ray or of other stars. But, to demonstrate that Moses was aware that the little lights were not only stars, he ended his sentence with "he also made the stars" (Ge 1:16). Once again, the following Bible verses can only amaze us :

Gen1 : 14 God said, Let there be lights in the expanse of the sky, to divide the day from the night; let them be signs to mark the eras, days and years;

Gen 1:16 God made the two great lights, the larger light to govern the day, and the smaller light to govern the night; he also made the stars.

Gen 1:19 So there was evening, and there was morning, the fourth day.

By this we see that Moses did not shirk scientific processes for the formation of the solar system, for scholars treat those who believe in the biblical creation story as people who lack arguments and who hide. Behind the phrase "God said, God made, God created". I had already said that the name of God is cited to demonstrate that the diagram was part of his will and his plan, but the processes were physical and I ask the scholars to listen to the above-mentioned documentary, recommended for this point.

On the third and fourth day of creation which is the second and third stage according to science, the Bible speaks of the emergence of the earth and the lights. Between stars and planets in the same galaxy, then what has formed before the other? It is up to the science of tomorrow to answer them exactly, according to observations made on earth from the 20th century, even if it appears today by calculation that it is the stars that were formed before.

3.1.1. Formation of the solar system

Astrophysicists will say that the sun was formed 9 billion years ago after the Big Bang, while the study of rocks shows that the earth was also formed at the same time. The Sun, the earth and the other planets of the solar system were all according to science, projected by the same star, thus confirming the verses which spoke of the waters below and the waters above coming from the same mass of water (Ge 1 : 6-7). If for scholars it was the accretion of material that was important in talking about the sun or the earth, for God or Moses it was the planetesimals, the seeds of dust of what will happen to the earth.

These planetesimals, called debris collected in the disc. These are dusts, which having gathered, ended up forming several coherent objects. When they were projected, what would become planet earth was already localized, detected in its orbit, and described as carrying life. The model on what originally happened we said, we will have to draw it by the phenomena we are witnessing. Speaking of the Big Bang, we will have to refer to the explosions of the stars.

According to science, by exploding in a supernova, a star simultaneously releases elements into space that will form new stars, or even new planets. As massive protostellar condensation develops around the center of gravity, which alone attracts most (99%) of the cloud material (mainly hydrogen and helium) that will form the star, the nebular disc, for its part, cools, thins and shrinks by taking an annular structure in which appears a condensation of denser particles (dust, crystals, ice, rocky bodies, etc.), in very varied orbits which will eventually form planetesimals.

In addition, these planetesimals agglomerate faster than the appearance of lights. However, it is the latter that will come to shape them into planets, asteroids or comets by their heat bombardment in general. Wait there until there are gaseous or telluric planets that escape this rule.

So, speaking of the creation of the earth on the third day by the heap of waters that lay below the expanse and of the lights on the fourth day, it is like saying that our oceans must have been older than the sun. This truth troubled me greatly until I came across this article, which careless scholars, because it does not fit with their models. This article appeared on leMonde.fr published on the Internet in m blogs on 09/26/2014 by Guillaume Cannat :

"By studying the origin of water in the Solar System, researchers have concluded that much of it was formed before the birth of the Sun and the Earth."

However, the formation of water takes place from nebular or protoplanetary discs. This conclusion is also consistent with the scientific documentary "Keys to the Universe 4 - The Origin of Water on Earth".

Of course, it is assumed that it looked like this :

The protoplanetary disk TW Hydra

(Cf:https://www.podcastscience.fm/dossiers/2019/09/30/la-naissance-des- planetes)

(Below is a diagram of the solar system today : the proto-sun became the Sun star and the planets formed around the Sun in the protoplanetary disk)

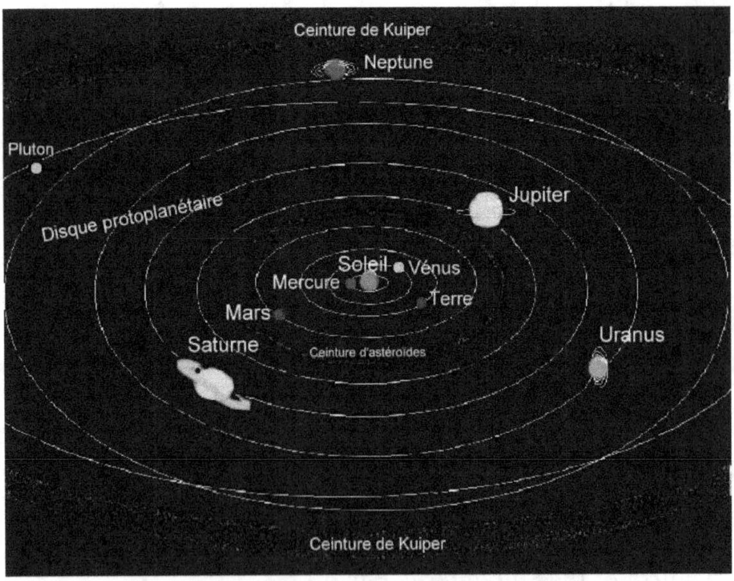

Image of the current solar system : the sun and its protoplanetary disk made up of planets and asteroid belts. (vinsvignesvignerons.com/Geologie/Geologie-de-la-France/ Une-breve- histoire-de-la-planete-terre).

3.1.2. Formation of other stars

While God focused on making the earth a viable planet after the Big Bang, what happened during those times in Heaven? For science it was the time of galactic collisions until stability returned. For us this stability returned when God, Our Father again thought of the water that was above the expanse as described by Moses, the scholar of scholars. (Gen 1:14-19)

Gen 1:14 God said : Let there be lights in the expanse of heaven, to divide the day from the night; let them be signs to mark the eras, days and years; 15 and let them be lights in the expanse of heaven to give light on the earth. And so it was. 16 God made the two great lights, the greater light to rule the day, and the smaller light to rule the night; he also made the stars. 17 God set them in the expanse of heaven to give light on the earth, 18 to rule over the day and over the night, and to separate the light from the darkness. God saw it was good. 19 So there was evening and there was morning, the fourth day.

Here, God - Our Father - has thought of everything according to life on earth. The lights should be aligned so that they not only disrupt life on earth, but contribute to the maintenance of that life. In speaking of the luminaries, Moses takes into account the solar system, whose large luminary represents the sun, the small luminary representing the moon and the stars representing the other stars which are outside the planet earth, because, when we look at the night in Heaven, everything shines like stars, including the planets which only reflect the light of the Sun like the moon.

This is how Moses suffered to make us understand in earthly and comprehensible terms the work of God. Frankly, if only for the scheme as set out in the first chapter of Genesis, even if one does not believe in God, the statue of Moses deserves the universal pantheon. I know that a lot of scientific work will be done from the publication of this book, but all the honors go to God - Our Father - and to his faithful prophet Moses who did not neglect any detail as if he knew that one day this scheme will explode like a bomb in the hand of Science.

3.2 Formation of stars by the same material

A star is understood to mean a luminous natural celestial body (stars) or not representing a physical reality. These bodies are found contained in galaxies and some are even isolated in space. What material were they formed by ?

Let's follow this verse by verse and note that everything comes from a state of material called "water" by Moses and which confirms that the Big Bang is biblical. For scholars, I point out to them that these verses are put here for scientific purposes and not as an act of evangelization and in addition, it is scientifically tested that these writings date from antiquity. Read :

Ge : 2 The earth was formless and empty : there was darkness on the face of the deep, and the spirit of God moved over the waters.

– God said, Let there be an expanse between the waters, and let it divide the waters from the waters.

– And God made the firmament, and he divided the waters which were under the firmament from the waters which were above the firmament. And it was so.

– God said, Let the waters which are below heaven be gathered together in one place, and let it be dry. And it was so.

– God called the dry land Earth, and he called the cluster of the waters Sea. God saw it was good.

14 God said, Let there be lights in the expanse (or empty space which separated the two waters) of heaven, to separate the day from the night; let them be signs to mark the eras, days and years; 15 and let them be lights in the expanse of heaven to give light on the earth. And it was so. 16 God made the two great lights, the greater light to govern the day, and the smaller light to govern the night; He also made the stars. 17 God set them in the expanse (empty space) of heaven to give light on the earth, 18 to govern the day and over the night, and to separate the light from the darkness. God saw it was good. 19 So there was evening and there was morning, the fourth day.

Returning to verse 7, we see that after the Big Bang the waters split in two. There was an expanse (separation or expansion) between the waters which were below the expanse where the earth was formed, and from verse 14 of the water which was found above the expanse there was the formation of what the Bible calls lights and stars.

But what caught our attention was that from the waters that were below the expanse the earth emerged. This means that the water and the earth coexist together, forming a Whole which is called planet earth (Ge 1 : 9 "that the waters which are below the Heaven gather in one place, and that the dry appears "). This development is very important for what will follow.

We spoke at length about the extent when we answered the question of what Heaven is. According to the Bible, expanse represents the empty space that encompasses everything, which is what the Bible calls Heaven. While the earth (the dry) appeared from the waters which were below (Ge 1 : 9).

Gen 1 : *9 God said, Let the waters which are below heaven be gathered together in one place, and let it be dry. And it was so.*

As for the lights, Moses said that they were placed not in the waters which were above, but in the expanse, that is, in empty space. This is how today we see that there are no static bodies in Heaven. Everything is moving. Note that the Earth is contained in the Sky, which is to say in the vacuum orbiting around the sun. This was already revealed in ancient times by the following verse :

Job 26 : 7 " He stretches the north over the void, He hangs the earth over nothingness."

We ourselves today, as laymen, looking at the stars, we think they are glued to the Blue Sky. But how could Moses or other Old Testament prophets accurately guess that all the lights were bathed in a void (expanse), and especially in their days ? To science, we have said, Moses was a myth and all the rest of the Old Testament was just mythology. Yes, if Moses was a myth, but the writings are real and were written by someone under divine inspiration. So we call what somebody said as Moses we said. Thus, it is no longer a myth. So, that science make him enter with Empedocles in its history just like Aristotle, Galileo, Newton, Einstein, Ptolemy and so many others.

In addition, science has always regarded these scientifically tested Biblical writings as statements made on the fly by their authors, and that coincidentally, this coincides with science. This book is demonstrating that these were thoughtful statements.

Moses, we have said, deserves the appellation of scholar of scholars and, this, by the way in which he describes the original material Pre Big-bang, having served for the formation of the earth and that this same material was used for the formation of luminaries, including stars and this is confirmed by the homogeneity of material throughout the Universe. What more can I add as proof that the biblical scheme of creation is scientifically approved ?

While Moses in writing these things saw only lights in Heaven, without being an astrophysicist, how could he have guessed or put together a story that the same elements that the earth is made of are also the same? that the elements of Heaven were formed? As the starting material, he called it waters. And the same water was used for the creation of all things as we can still read it again :

Ge 1 : 2 The earth was formless and empty : there was darkness over the face of the deep, and the spirit of God moved over the waters. 6 God said, Let there be an expanse between the waters, and let it divide the waters from the waters.

Nowadays, had it not been for the study of material in the Universe, no one could believe this pattern. This is why until then, these Bible verses of the first chapter of Genesis have never been the subject of much exploitation neither by scientists, nor by religious because Moses used the word "waters". The religious were afraid to focus on the first chapter of Genesis so as not to fall into atheism. Now, in these days, we are discovering that if there is one story that can bring us closer to God, which makes God a living reality, it is the pattern of creation contained in the Bible and not the miracles that even the Devil can do. Besides, Didn't he done so by telling Adam and Eve that they were going to know good and evil as God and that it was accomplished?

The verses (Ge 1 : 7 and 14) have always gone unnoticed, but, make God the Creator of the Universe. Moses, the scholar of scholars, did not have a giant, high-performance telescope to affirm things that science began to discover in the 20th Century, namely the Big Bang and the homogeneity of material in the universe. Read :

Ge 1 : 7 And God made the firmament, and he divided the waters which were under the firmament from the waters which were above the firmament. And so it was.

Gen 1:14 God said : Let there be lights in the expanse (or the empty space which separated the two waters) of Heaven, to separate the day from the night; let them be signs to mark eras, days and years.

The incomprehension that was between scholars and religious and to which the book "The Verses of Scientists" has just lifted the veil was due to language. Indeed, for an English speaker who does not know French, the name John is different from that of Jean and for a French speaker who does not know English or Russian, Michel is different from Mikhail or Michael. Only someone who is at the center of two languages can understand that John and John are the same name as are Michael, Mikhail and Michael.

Rather than attacking scientists like all religious have done until now, I preferred to put myself in the center by trying to study the two languages and I discovered that even if the names or the words were different in the story of creation, but we spoke the same language. The understanding between the words used in scientific language and in biblical language was the same.

Another example; when I found out that the biblical abyss means the black hole, I had felt miserable for 3 days and thought to myself, I will never publish this book or I will be ridiculed by the scientific world and behind me, a whole innocent organization.

Indeed, as we said before, for the scientific world, the black hole was formed after the Big bang and yet the Bible tells us that the abyss (black hole) existed before the separation of the waters from the waters (Big-bang). Thank goodness I came across an article speculating that the Universe may have come from a black hole :

"It's quite tempting to imagine that our Big Bang could end up coming out of a wormhole. This is one of the few hypotheses on which researchers are currently working".

(Cf :05/11/2014 - By Jonathan SARE, Futura-Sciences).

And, later, having known the significance deformation of the fabric of space-time, which is only an abyss in space caused by the presence of a mass, the biblical scheme of creation became for me a scientific reality. In an empty universe, when the primordial gravitational force or Universal Prime Force began to bring together the original particles, described by the biblical phrase "'a wind of God moving on the waters", it distorts the tissues of the space-time described by the biblical phrase "there was darkness in the face of the abyss".

As we can see, the knowledge about the source of the Universe is found in the Bible and as I said, astrophysicists have no choice. Entering the Big Bang Front brings you closer to the cause which is God (the spirit of God was moving over the waters). And for this, knowledge of the Divine Principle of the Moon couple is essential. This knowledge was obscured by the media image of him.

I'm sure NASA is only a small kitchen compared to Moses' laboratory, and today's 20th century satellites are still archaic to what Moses used. The latter was able to see into the heart of the last galaxy by entering the heart of God. To come to the conclusion that it all started with a single whole which he called "waters" and then exploded after thousands of years, it was necessary without a laboratory and a refracting telescope to witness these events.

Belgian Abbot Georges Lemaitre came to this conclusion after observations of the remoteness of galaxies observed by the telescope of the American Edwin Hubble. Despite that, he didn't say anything about the Pre Big Bang because he didn't take a look in the Bible and consider analyzing it by comparing it with his discovery.

In the Universal Cosmological Model, the chance of

nature does not exist. But that "the will of God and the chance of nature" do not become a source of division. What can be a source of division is when scientific laws or arguments about the formation of the Universe contradict each other. For example, in the Universal Cosmological Model, dark energy (darkness) preceded the Big Bang as well as one of the cosmic rays (light).

Thus, the great biblical radiations, darkness (dark energy) and light (cosmic or fossil radiation), attested and discovered by science, continue to accompany us to this day. If cosmic or fossil radiation were a consequence of the Big Bang, each exploding star should contribute to its increase, because the scenario is the same. But, as the Bible says, "from the spirit of God (energies) which moving on the waters (the elements Pre Big-bang), light comes forth" (Ge 1 : 2-3). And that took place before the Big Bang. So the decreasing temperature of the fossil radiation proves that it was a pre-Big Bang light as the Bible attests.

Section 4 :
The Axis of the Universe before and after the Big Bang

4.1. Biblical illustration of the universe before the Big-bang

Ge 1 : 2 "The earth was formless and empty; there was darkness on the face of the deep, and the Spirit of God moved over the waters."

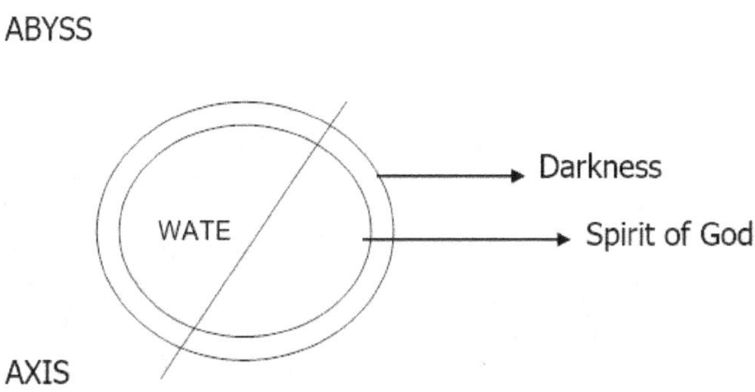

4.2 Biblical illustration of the universe after the Big Bang

Gen 1 : 7 "And God made the firmament, and he divided the waters which were under the firmament from the waters which were above the firmament. And it was so".

Separation of water from water or Big bang.

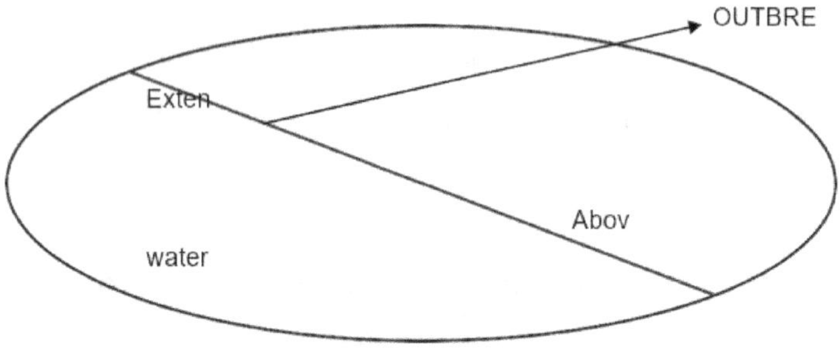

Disappearance of the axis preventing even the creation of multiverse

We later discovered images showing our prediction that the axis would disappear after the Big Bang. Here are the most recent :

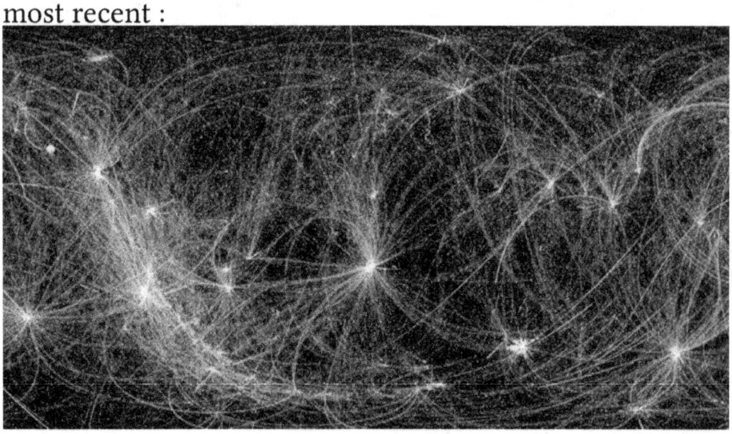

This image of the sky was produced after 22 months of observation from the ISS using the NICER instrument, targeting all cosmic sources of X-rays.

(Https://trustmyscience.com/wp-content/uploads/2019/06/carte-rayonsx- ciel.jpeg)

Another scientist also expressed the following point of view through this illustration :

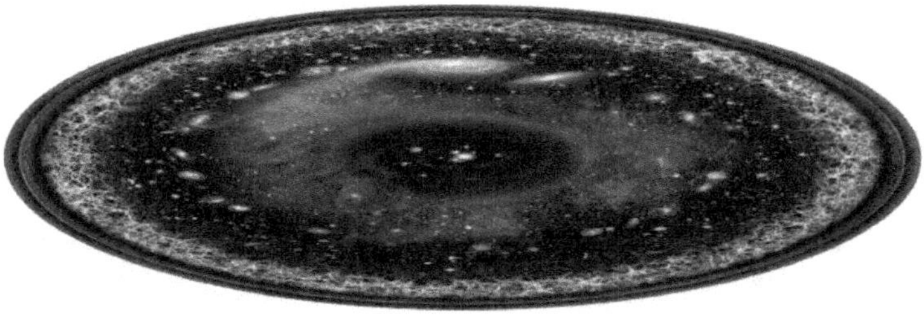

"The Big Bang did not take place at one point, but simultaneously in all directions. And if the Universe does have a center, the observable region of the Universe that we see is insufficient to give us this information. So virtually any observer can be the center of the Universe, as in this logarithmic scale representation of the observable Universe, centered on the Solar System". Credit : Pablo (Cf:https://trustmyscience.com/nouvelle-etude-matiere-noire-plus-ancienne- que-big-bang/by Stéphanie Schmidt, August 9, 2019).

Chapter VI : The vegetable Kingdom, beings Aquatic and ruffles

(5th day)

Section 1 : The plant kingdom

1.1. The evolution of the plant kingdom

Let's go back first to the third day of creation where we find these verses :

Ge 1:11 Then God said, Let the earth bring forth greenery, and grass bearing seed, and fruit trees bearing fruit after their kind, and having their seed in them on the earth. And it was so.... 13 So there was evening and there was morning, the third day.

Following these lines of reasoning, which Moses did not even know that one day would be analyzed, everything starts from the infinitely small to the infinitely large : greenery - grass - fruit tree. We see that the whole process is detailed in the Bible in a logical way. Scientifically, the process of life appearances is very complex. But, the most important point even in science is that in materials of creation everything starts from the infinitely small to the infinitely large.

We have already said that the third day of creation corresponds to the second age of science and it is a period of galactic and lifeless collision. Moses spoke of the plants in the planet earth to express the presence of the molecules of life in this cluster of dust as described in the scientific documentary "the origins of the earth"; written by Alice HARPER and Nigel HENBEST.

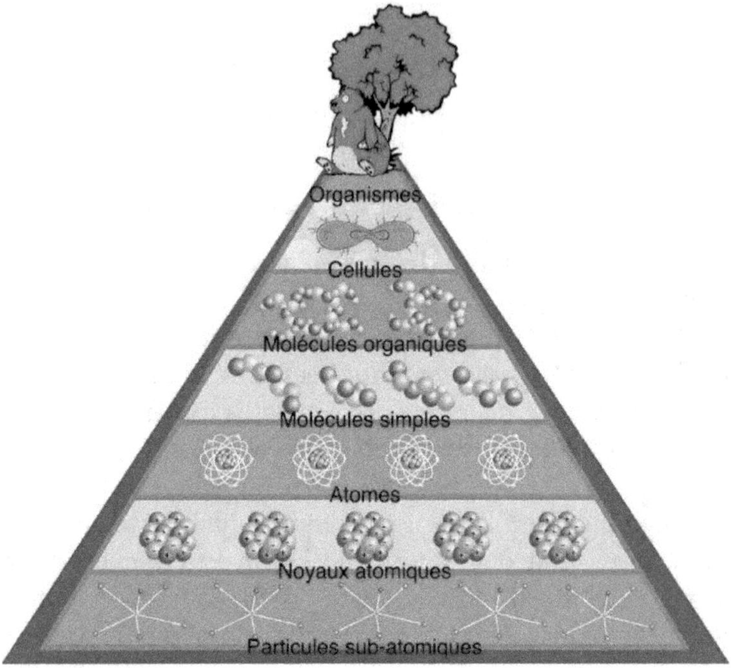

astrocanada.ca/le_cosmos_contemporain_la_matierecontemporary_cosmos_matter/Fra

Indeed, after the Big Bang and the other processes that followed, organic molecules had redistributed in the Universe. Those that were on planet earth, with the environmental combination created by its atmosphere, should give the life that Moses described. The most striking example we can draw from the appearance of plants in the Galapagos Islands.

The Galápagos Islands were formed by submarine volcanoes. Originally, these islands were completely barren, with no life forms of any kind. Scientists attribute the appearance of the greenery to the supply of birds that came to shelter there by bringing seeds. Whether argument founded or not, Charles DARWIN who also visited this island, concluded in his theory of natural selection or the resistance of the fittest, that the seeds were brought by sea currents, even going so far as to apply the theory of spontaneous generation as phenomena for which no explanations could be given.

We don't deny an answer, but not all of the greens there are just seeds. This is to say that the particles carrying life can be found in environments that do not allow them to evolve as in other planets or comets, and this for lack of the ideal environmental conditions enjoyed by the earth. Such particles will always remain primitive and inert, because they are bombarded either by rays from their stars, or because of very extreme climatic conditions that prevail there, not causing the presence of liquid water.

On the other hand, in Galapagos Island, created according to science by volcanic magmas, the molecules which had buried there in depth found themselves in the open air and were able to benefit from the ideal conditions which gave them chance to grow in plants.

To return to the appearance of plants, Moses did not take into account the universe in general, but focused on the appearance of life on earth on the 5th day and not on the 3rd day (Ge 1 : 11 - 13) which only concerned the molecules of life as we can read in Genesis 2 from verse 4. Read :

Ge 2 : 4 These are the origins of the heavens and the earth when they were created. When the Lord God made the earth and the heavens, 5 no shrub of the field was yet on the earth, and no grass of the field yet sprung up : for the Lord God had not rained on the earth, and there was no man to cultivate the soil. 6But a mist rose up from the earth and watered the whole face of the ground. 7The LORD God made man out of the dust of the earth ; he breathed into his nostrils the breath of life, and man became a living creature.

As we said, planktons which are among the first living things, made up of tiny plants and animals, fall into two categories. The smallest is phytoplankton, plant plankton. The other category is animal plankton known as zooplankton. Zooplankton feed on phytoplankton. This confirms that it is the plants that were created before the animals, as the bible says.

In these verses, Moses is giving us a scientific demonstration of how things turned out, which indeed fits in with science as we will find out in the following points.

1.2. Water and life in the plant kingdom

Ge 2 : 4 "... When the Lord God made the earth and the heavens";

Ge 2 : 5 "... and no grass of the field was yet sprouting : for the Lord God had not made it rain on the earth ...".

Here, Moses focused directly on the planet earth. Indeed, the chemical composition of water had to reach its current level to ensure and sustain life. We also talked about this before. But we're coming back because some readers will find the points on the Pre Big-bang and the Big-bang very complicated, and tend to start there. Let's continue our reading of Bible verses :

Ge 2 : 6 *But a mist rose up from the earth, and watered the whole face of the ground*

In Ge 2 : 5, we are told that no grass was sprouting, for there was not yet the rain. Then in verse 6 we are told that the earth was watered. This shows that it was after the fourth biblical day that God created lights (Ge 1 : 14-19). Which is to say that the first plant appeared in the 5th stage as evidenced by science in the table at the beginning. This is confirmed by the following verse :

Gen 2 : 7 "The LORD God made man out of the dust of the earth"

Now, man was created in the sixth step. So the plants according to the above verses were formed in the fifth step. But for Moses the creation of the earth and of the life forms a whole and that started with the unicellular algae of which he could only describe them by plants. Moreover, according to various scientific sources, these algae (cyanophyceae), some mutated to plants ; phytoplankton and the others in marine beings ; zooplankton.

In principle, all sea grasses are algae. They are not often considered true plants because they make their own food and lack roots, stems, leaves and flowers.

In addition, when God created man, he said to him :

"Behold, I give you all the grass that is seeded, and that is on the face of all the earth, and every tree that has the fruit of a tree in it, and that bears the seed : it shall be your food. 30 And to every animal of the earth, and to every bird of heaven, and to everything that moves on the earth, having within itself a breath of life, I give all green grass for food" (Ge 1 : 28).

This proves that the creation of plants, that is to say algae, to precede animals and man.

According to science, in the early or Precambrian era, at 4.5 billion years, there was formation or appearance of the earth. 500 million years later, or 4 billion years later, it formed the primitive sea. There was still a great period of one billion years for the oceans to reach chemical proportions for the activation of the molecules of life (Ge 2 : 6).

Just as Moses closed the Pre Big-bang with the appearance of the light, demonstrating that the warming of the primitive materials had reached its peak before exploding, here he used the coming of the rain before the appearance of the life to demonstrate that the earth's environment could now allow life to flourish from the level of chemistry that water has reached. This was confirmed in this quote from the science documentary *"Cosmos - beyond the Big Bang"* written by Matt HICKEY : *"Underwater, mysterious chemical reactions come to create life"*. A phrase that I really like, uttered by a scientist and demonstrating the limitation of science when one approaches the physics of creation ; branch of Astrotheology integrating God, the infinitely small, the Principle of creation of Rev. Dr. Sun Myung Moon and quantum mechanics.

Note that the Principle of Creation is the part which answers "to the why" that science cannot answer and which will propel last towards level top for the physical realization of the kingdom of heaven on earth as it is. says in this verse :

Isa 65:17 For I will create new heavens and a new earth; Past things will no longer be remembered, they will never come to mind.

Isa 65:20 Never again will there be in it an infant who lives but a few days, or an old man who does not live out his years; the one who dies at a hundred will be thought a mere child; the one who fails to reach a hundred will be considered accursed..

This is how religion must have predicted since ancient times, when scientific developments will peak. Namely, the prophecies of Isaiah, date from the 8th century BC. J.-C.

1.3. The protection of the plant kingdom

1.3.1. The message of plants

The man being part of the weak swimsuits of mammals, one wonders how he was not extinct in the face of all the microbes that surround us, some of which are very ruthless, during his two million years of existence. We will owe this magic only to the plants that God has made available to us.

But, scientists want to seek the answer further, by doing what is biblically forbidden. They seek to resuscitate Neanderthal man by cloning, extinct 30,000 years ago and whose cells are thought to be found one day alive through bones buried either in a glacier or in a humid underground environment. They want to understand how his body was immune to germs in order to include the treatment in our genes, they say to skeptics. We will come back to this when we discuss cloning in the following pages.

Each plant has a message that it conveys to humans for their happiness, by which they mean their health. This is written naturally in these sheets. So many species that have disappeared to these days represent many messages that we have lost, when we could have exploited it. Now is the time for green movements to raise the tone for the protection and good use of what we have left.

For us believers, God has provided everything for plants to meet the needs of the body, whatever the season. It is often said that each change of the seasons brings about new diseases. All this is because Humankind no longer knows how to take advantage of what nature offers him.

Indeed, with each change of seasons a series of fruits appear in abundance to allow the body to cope with the inconvenience they bring. For example, if there are more citrus fruits during the cold season, it is because the body needs more vitamin C during this time. Many people get sick during this time because the lack of vitamin C lowers antibodies and allows germs to grow. It will also be found that the fruits that appear during the hot seasons are richer in water than the fruits of the cold seasons because it is hot at that time, which allows the body to avoid dehydration for those who consume it.

For those who want to know more about the fruits to consume during the different seasons, I send them on these links *https://www.fruits-legumes.org/calendrier-fruits-legume*, even if it means making the choices according to the environments. For example, a hot season in Europe can correspond to a cold season in Africa. So, you have to know what season you are in your environment and see the fruits expected for these kinds of climates.

1.3.2. Location of the most potent medicinal plants

What is the place of plants in the life of humans and animals ? God, the Creator of all, answers :

Ge 1:29 And God said, Behold, I have given you all the herb that is seeded, and that is on the face of all the earth, and every tree that has the fruit of a tree in it, and that bears seed : it shall be your food 30 And to every animal of the earth, to every bird of the sky, and to everything that moves on the earth, having within itself a breath of life, I give all green grass for food.

Ps 104 : 14 He causes grass to spring up for cattle, and plants for human needs, that the earth may produce food,

Finally, without leaving science to trial and error, God gives a precise indication for the health of man by telling him that all the leaves of a fruit tree, growing on the banks of a torrent and whose fruit growth does not break, are a very powerful remedy. Against what ? It is for science to find out by studying these molecules. And even their fruits should be used for immunity against certain bacteria and viruses. The ball is now in the camps of science, that is, in the hands of man to explore all that God has provided for him and to thank Him. As mentioned in the verse below, the study of the molecules of the leaves of fruit trees that grow in the vicinity of torrents is necessary as we will read :

Ez 47:12 All kinds of fruit trees will grow on the brook, on its banks on either side. Their leaves will not wither, and their fruit will never end; they will ripen every month, because the waters will come out of the sanctuary. Their fruits will be food, and their leaves medicine.

This as well as during the CORONA VIRUS pandemic which grieves humanity when we submit this book for editing, President Andry Rajoelina of Madagascar, announced on 04/20/2020 the discovery of a miracle product helping to prevent covid-19. It is an organic herbal tea - civid organics - made from artemisia, a plant whose active principle is known to treat malaria, mixed with other plants not mentioned. Namely that Madagascar, which is a country surrounded by water, with a rich ecosystem offers conditions allowing to find such plants.

While waiting for research to prove its effectiveness, what interest us in our message is the importance of plants, as well as this indication of the Prophet Ezekiel who lived in the 6th century BC. J.C.

Section 2 : Aquatic beings and sea animals

2.1. Aquatic beings

In the scientific documentary *"The Origins of the Earth"* written by Alice Harper and Nigel Henbest, we are told that the appearance of fish was followed by flying insects. The Bible, speaking only of purpose, mentions winged beings instead, demonstrating that science has reached it. Let's read :

Genesis 1 : *"20 God said, 'Let the waters swarm with living creatures, and let birds fly over the earth toward the expanse of heaven. 21 And God created large sea- creatures and every living creature that moves, and swarms in the water, according to their kinds, and every winged bird after its kind. God saw that it was good. 22 God blessed them, saying, "This is good : Be fruitful, and multiply, and fill the waters of the seas, and let the birds multiply on the earth. 23 And there was evening and them morning : the fifth day".*

Until then, science has focused much more on the appearance of the animal kingdom, forgetting that the latter, even the archaeobacteria, cannot find refuge or have a give and receive ensuring its survival, without the presence of vegetation. This is how it can be found even at the bottom of the sea. At the last information (2017), the age of emergence of plants has been further reduced 400 million years ago.

Previously, the assumption was that without the sun there is no life. But let's see this article taken from the "energy" review no. 4 May 2004 published by Total :

"Over the past thirty years, submarine exploitation has completely called into question our vision of evolution on

earth, from the conception of the continents, to the conditions under which life appears on our planet ... By exploring a ridge in 1977, off the Galapagos, American scientists were surprised to find geysers of very hot black water (between 150 and 400 ° C) at less than 2600 m ... Around these geysers, in an environment that 'it was believed to be totally deserted, an ecosystem completely independent of life on earth had developed. Exuberant fauna : worms almost 2 m long, giant bivalves, crustaceans' unknown until then. With this discovery fell a myth. Life believed to be closely linked to photosynthesis was also possible outside of the sun's action. Because at the origin of these underwater "oases of life" are microorganisms, archaebacterial (first living organisms to appear on Earth according to science) which flourish in these extreme conditions, close to the environment of the earth at its origins, between hydrogens, methane's and heavy metals. We then speak of chemosynthesis".

Thus, we notice that the animal kingdom either derives its food from plants, and either inspires the gas that the latter emit, or both at the same time like man. Thus, the biblical logic of creation is an absolute and indisputable truth. But by mixing Genesis 1 : 23 and Genesis 2 : 4-6, we conclude that plants and sea animals from algae all appeared on the fifth day of creation according to the Bible and in the fourth era according to science.

2.2. Flying beings

According to the Bible, another significant point is the appearance of fish, followed by the appearance of birds. Indeed, this is confirmed by the existence of water or aquatic birds, including the emperor penguin, which dives more than 500 m in search of food :

Ge 1:20 God said : Let the waters bring forth living animals in abundance, and let birds fly over the earth to the expanse of heaven.

Some scientists will say that by jumping above water to inhale air, some fish have developed wings, becoming birds, of which the penguins serve as hyphens.

But let's see how a biochemist explains this transition :

"The first multi-cellular life forms were algae and sponges, followed by jellyfish, worms and starfish. The first animals with a backbone were fish. Some of them evolved into amphibians. These gave birth to reptiles (crocodiles, snakes and lizards). The largest among them was the dinosaur. Some small dinosaurs adorned themselves with feathers and evolved into birds, while another group of dinosaurs gave birth to mammals, which to this day remain the most evolved animals on the planet".(Cf : Linda Gamlin, The Origins of Life, ed. Artis-Historia, Brussels. 1989, P.3).

Here we have the scientific scheme that led from underwater beings to birds, whereas the Bible only speaks of very important stages that are understandable to everyone. Speaking of underwater beings, it includes the evolutionary stage from the first forms of underwater life to birds, the last species of flying beings to appear.

For example, how could Moses talk about certain beings that people had no knowledge of? So, for the Universal Cosmological Model, every finite and non-transitory form was the culmination of a principal.

Chapter VII :
From mammals to humankind

(6th day)

Section 1 :
The cohesion of the scheme of biblical creation

In our leading scientific documentary for this point, "The Origins of the Earth" written by Alice Harper and Nigel Henbest, we are told that between fish and reptiles, there were first flying beings. This also fits in with the biblical verse where reptiles are not found on the 5th biblical day like fish and flying creatures, but on the 6th day (Gen 1.25).

But which reptiles are we talking about here, since in the quotation below, from the biochemist and biologist Linda Gamlin, reptiles that evolved into dinosaurs are said to be the ancestors of birds ? Does the Bible contradict science ? Not at all, because in the last sentence of the same quotation we find another category of reptiles (dinosaurs) as we can read :

"*Some small dinosaurs adorned themselves with feathers and evolved into birds, while another group of dinosaurs gave birth to mammals, which to this day remain the most evolved animals on the Planet*". (Cf : Linda Gamlin, The Origins of Life, ed. Artis-Historia, Brussels. 1989, P.3).

In all this, we discover that the pattern of the appearance of things or beings in broad outline is almost in harmony between science and these writings of 1500 BC. The purpose of Moses, in writing the creation story, was to make the Israelites see that God is the master and Creator of all things, in as much as it is he who created the original atom "waters" as scientifically attested :

"*Today, there are at least 4 million known species of living organisms on earth. They all use the same genetic code and their basic chemical composition is identical, which seems to indicate that they all descend from the same primitive species of bacteria, which were alive more than 3 billion years ago".* (Cf : Linda Gamin, Les origines de la vie, ed. Artis-Historia, Brussels. 1989, P.24).

I found the answer to the question that led me to write this book, namely to verify whether the chronological order of creation according to science was different from that of the Bible. At 95%, it's the same pattern. Since there is research that needs to be done between certain contradictions, it is up to science to confirm the remaining 5%. For us believers, the biblical order as rendering in the Universal Cosmological Model is 100% true. Since this book reassures scholars of the truthfulness of the biblical creation accounts, it is up to them to begin reviewing their calculations in order to achieve divine order.

It should also be pointed out that the Standard Scientific Model does not detail all these points. Hence the Universal Cosmological Model, which is richer in opening up research from before the Big Bang to man.

Let's go back to our material and see what happened on the 6th biblical day or the 5th era according to science :

Ge 1 : *"24 God said, "Let the earth produce living creatures according to their kinds : livestock, creatures that crawl, and the wildlife of the earth according to their kinds... And it was so. 25 And God made the wildlife of the earth after their kind, and the livestock after their kind, and all the creatures that crawl on the ground according to their kinds. And God saw that it was good.26 And God said, Let us make man in our image, after our likeness, and let him have dominion over the fish of the sea, and over the birds of the sky, and over the livestock, and over all the earth, and over every creeping thing that crawl on the ground. 27 God created man in his own image, in the image of God he created him; male and female he created them...31 And God saw all that he had*

In the sixth period of creation, called the sixth biblical day, we see here the creation of the beings which are very close to us, namely the earthly animals.

If the Bible was written at the time when man was discovering, it would have been said to have been the fruit of human calculation, based on the dating of fossils. In addition, today, we have a specialization of intelligence so that the biblical creation story should call upon several scientific disciplines for its cohesion.

Even with an intelligence specialization today, the mystery of creation still resides. Now, in the story of Genesis, the same person "Moses" was able to talk about everything : from before the Big bang to the appearance of Humankind in a consistent manner. If anyone still finds an inconsistency between the schema of biblical creation and science, they haven't read this book carefully. It is up to science these days to give an attribute or a description to this designer of the story of Genesis chapters one and two, if not that of scholar of scholars.

Section 2 :
The Structure of Living Things

In Ge 2:7 we read that man and animals were formed out of the dust of the earth (Universe), which science could never deny, because by analyzing the dust of comets, the molecules of life are constantly being discovered. But these molecules cannot interact with each other without liquid water with chemical proportions such as the water that is on planet earth.

Thus, before the creation of humans and animals, Moses began by first mentioning the problem of water.

In Gen 2:5" no shrub of the field had yet grown on the land, [d] and no plant of the field had yet sprouted, for the Lord God had not made it rain on the land, and there was no man to work the ground. 6 But mist would come up from the earth and water all the ground. "

Gen 2:7 "*The LORD God formed man from the dust of the earth*".

Even the Prophet Muhammad (peace and salvation be upon him) said that man was formed from mud, meaning the contribution of water in the dust. Once again, we discover a unique source of inspiration between science and religion, namely God.

So, even when science uses the terms molecules of life, it is well aware that these have nebulae as their primary source. However, in the nebula, there are gases and dust. It is dust that will one-day form the molecules of life in stars according to science. And, it is in the dust collected from comets that we also begin to find these molecules of life.

Thus, Moses will use the source to speak of dust as an element that was used in the creation of life, starting with plants.

Again we see that science and the Bible speak the same language. But, it was the Bible that predicted before science came to find them. To disorientate this discovery by bypassing the Bible, scientists preferred to speak of man, as being created from star dust.

2.1. The dust of the earth

Ge 2:7 "*And the LORD God formed dust from the ground, and breathed into his nostrils the breath of life, and man became a living being.* "

According to astrophysicists, man is formed from star dust. Indeed, like all matters, the process of the creation of man,

for scientists, began with the fusion of particles until they formed hydrogen at the beginning of the Universe. Hydrogen, upon heating, eventually decomposed into helium. The repetition of this type of process - allows the creation of the x's of the different chemical elements known up to iron.

All these elements, thus created, are found in the form of fluids and plasmas and are held by gravitation under a ball that we call the star. Iron is generally considered to be the final stage in the chain of nuclear fusion reactions within stars.

Thus, at this stage, fusion cannot continue to evacuate its energy by building heavier elements. The ball then begins to absorb its energy instead of emitting it. And that leads to its implosion. Once the star or ball explodes, the increase in temperature it gives off creates yet other heavier elements.

By exploding, through a phenomenon called supernova, the outer layers propel into space elements they have made inside them and gases we find in nature, including molecules of life.

The 99% of this material has come together again by the same starting process and will form the Sun. Other debris found in space far from any fusion process to initiate thermonuclear reactions will also begin to assemble by the force of gravity. These debris are called planetsimals. These are dusts, which having grouped together, eventually form several coherent objects in their orbits of projections, forming planets, including the earth. This is why in each planet we find the same elements as those that make up the stars, confirming the Big Bang or the separation of the waters from the waters.

Finally, it is in the planet Earth that man will eventually be formed by the elements that make it up, which Moses called the dust of the earth. Scientists referring to the origin, i.e. the star ball, say that man is made of star dust. The Prophet Muhammad (peace and salvation be upon him), knowing that the peculiarity of life on planet Earth was due to the presence of liquid water, will say that man is made of mud ; that is to say, the dust of the earth plus water. And, the last great prophet to think about this material, namely the Rev. Dr. Sun Myung Moon, concluded, in the Divine Principle, that man is the microcosm of the cosmos.

Knowing that it is God who led all the processes from the origin to the formation of man as the last stage of creation, man finds himself to be the son of God and the Lord of the Universe. (Gen 1:27).

2.2 The breath of life in living beings

In the process of evolution, the difference between other living beings and man came from the divine aspect, or the consciousness attributed to each being, defined as the breath of life, an unfathomable field for science. Let us read

Gen 2 : 7 And the LORD God formed man out of the dust of the earth; he breathed into his nostrils the breath of life, and man became a living creature.

Until then, the believers, in their literal analysis, thought that the breath of life was equal to the air. However, air is physical; it is the name given to the earth's atmosphere, which is composed of several gases. The breath of life, here, is part of something spiritual that cannot be felt. It is in fact the energy that is endowed to each being, as we have said, starting with the original particles that connect us with God. It is also the source of consciousness.

This verse cannot be interpreted as if God has breathed the air we breathe into our nostrils. This ignorance resulted in death being assimilated by man's inability to stop breathing. With the development of medicine, we understand that the two primordial organs for sustaining life are the heart and the lung and not the latter alone. It is through these two organs that we can revive an animal being.

What God has blown into man has nothing to do with physical air. This breath has remained until today an enigma for those who wondered about things that cannot be felt in an individual, mainly the spirit and the soul. It is also through this that consciousness entered into inert material.

Moses describes it as the breath of life which is energy from God that the Moon couple explains in the book of the Principle of Creation and which is called the Universal Prime Energy and which was endowed to every living being and man was used as a case in point.

By saying that man was inert, and that the breath of God gave him life, we find ourselves in a phenomenon that scholars can describe as "the entry of energy into the human being". In the Divine Principle of the Moon couple, who are the first to explain the power of God in a scientific way, this sentence is articulated as follows :

"The First Universal Force has its origin in God and constitutes the force that God projects in each being at the moment of his creation" (Cf. : The Divine Principle, level 4, P.21).

In the breath of life, we find the spirit and the primary universal energy coming from God, and transmitted to living beings. The notion of Universal Prime Energy and Universal Prime Force are new to science and theology. This book, being much more descriptive, is not suitable for giving detailed explanations. The teaching manual for these concepts is the Divine Principle of Rev. Dr. Sun Myung Moon.

As I keep saying in my textbooks, biblical language is spiritual; the physical term is used to teach believers. But the real explanation is to get into the heart of the one who said it.

Having exposed all this, we define the Universal Prime Energy, or the breath of life, as the energy of life. Without this energy, life becomes inert. This energy did not begin with the first molecule, but long before the Big-bang. It comes from God Himself. When we see a living being, we feel that it has energy, we feel that it has life, but we don't see this life sustaining energy. Its manifestation within the being creates the First Universal Force.

Chapter VIII : Humans and the Universe

Section 1 :
The Responsibility of Man in the Universe

Gen 1:28 "God blessed them, and God said unto them, Be fruitful, and multiply, and fill the earth, and subjugate it; and rule over the fish of the sea, over the birds of the sky, and over every animal that moves on the earth".

Cor 6 : 3 "Do you not know that we will judge the angels? And all the more reason we would not judge the things of this life"?

Through these verses we see that man has been established as master and lord of the Universe, whether over things visible and invisible. The term "dominated" here implies the word "reign". At that, we will have to understand our heavy responsibility in different areas :

1.1. Cosmology

In the cosmological field, we are called to discover our Universe and to know everything that surrounds us, their impact on us and, above all, what we can benefit from. These are not things that were created for our contemplation. The effort should be shared and everyone must make their contribution.

In the Divine Principle of Rev. Dr. S.M. Moon, we read this :

"Through science and technology, man is restoring his external dominance over creation. Vast scientific developments have taken place during this century. Through these advances, man has been able to control disease, oceans, land, air and space. "

All of this falls within the scope of God's third blessing to man, namely, "... rule the universe..." (Ge 1:28).

1.2. Environment

In the environmental field, we will have to discover and know what is closest to us. Protect species of animals and plants that have not yet been exterminated. Encourage the construction of skyscrapers and HLM (low-rent housing) to promote vertical growth of the area rather than horizontal growth which reduces space and therefore very quickly destroys the ecosystem. This point also fits with the divine recommendation as we can read it :

Isa 5 : 8 "Woe to those who add house to house, and join field to field, until there is no more space, and they dwell alone in the midst of the land."

Moreover, with the demographic boom, we risk destroying the entire ecosystem by conquering uninhabited spaces (devastation of forests). It is in our hands that lays the power God has given us, to destroy or build. But we will have to build and protect.

Section 2 :
The air

2.1. The importance of physical air

As we said, air is physical. It is a gaseous fluid that makes up the atmosphere. It is essential not only for the life of organic beings, because it participates in the process of respiration, but also to plant in photosynthesis. It is made up of 21% oxygen, 78% nitrogen and 1% of rare gases (mainly argon).

The conditions with which the earth is placed in the cosmos are very exceptional. We will have to be aware of this and be jealous. The race against wealth is weakening its self-protection, notably the exploitation of raw materials and the release of gases into the atmosphere which increases the greenhouse effect, thus accelerating climate change.

We will have to learn the lesson from the extinction of the dinosaurs according to scientific hypotheses, because man by living in cloistered spaces, has become the weakest being to resist a sudden change of climates. A sharp rise or fall in this would lead to the extinction of living things on a large scale. Hence the strengthening of movements and ecological measures for its protection

Because of pollution, the body of people in the future should also adapt to the integration of carbon dioxide (or carbon dioxide). At the moment, its content in the air is very low (about 0.03%), but it keeps increasing at the rate of 0.5%. per year, thus actively contributing to the greenhouse effect. (Cf. : "air." Microsoft® Encarta® 2009 [DVD]. Microsoft Corporation, 2008).

In addition, the air we breathe is found in the universe.

Since it is an integral part of our life, through it we become an extension of the universe. It is like an electric wire from the universe to us ; the Universe being the generator. On earth, this air has appropriate chemical properties adaptable to the beings that reside there.

The breath of life, on the other hand, is spiritual and energetic. It is a kind of projection of the spirit of God on us so that we have life spiritually and physically as we can read :

Jn 1 : 4 In him was life, and the life was the breath or the spirit of God in created beings.

2.2 Contribution of air to our organs

As blood circulates through our veins, the air is like a current wire that is connected to us and of which the universe is the generator, we said. Living things on earth cannot dissociate from the air, so with the Universe, we are one whole or all one. If man was created on the moon, his body couldn't have lungs, because there is no air there. Scholars do not like such examples which prove the existence of an intelligent being who reasoned to create us, namely God.

This relationship between human organs and the air makes us the universe ourselves. This is what causes us to make no effort to breathe and sometimes we are not even aware of it. By this, we understand that air has the same value as blood and that we will have to watch and purify what enters us, even if we make no effort. If not, even breathing unconsciously should cause a man to lose energy and his lifespan will be reduced.

People wait to talk about the greenhouse effect and think it's politicians' business. The greenhouse effect that results in increased heat is just a thermometer of the pollution that assails us. Those who don't understand this and shun environmental NGOs have yet to learn lessons. Sacrificing even nobody's future for economic gain is a crime against humanity.

Air is part of our component. It only takes a few minutes of deprivation for us to be asphyxiated. We talk more about the greenhouse effect than about environmental pollution, and people do not feel directly concerned.

It is time to start teaching all students not only what food brings in our bodies, but also the air. What we will also have to do to keep it pure in our environment. Until then, breathing is only taught in the broad sense : a release of carbon dioxide (CO2) and, or absorption of oxygen (O2).

By the way, we can say that breathing air is a vital phenomenon whose purpose is to oxygenate the blood. According to several medical sources, the heart of an adult at rest pumps about 4.5 L of blood per minute. These 4.5 L, carry 1.5 g of oxygen. For water to be the only source of this oxygen in our blood, we will all need to be reduced to the primary level of aquatic beings. In addition, we will need to have an organism ready to simultaneously collect from the water the oxygen contained in it and transfer it directly to the blood as fish do.

This is because fish breathe oxygen just like humans. But, they shoot them in the water and not in the air. Their fins separate oxygen from water and their blood carries this oxygen throughout the fish's body.

Knowing that living things on earth will evolve outside of water, God has designed a very good thing for us. Equipping us with lungs for breathing, rich or poor, brave or lazy, the air has been spared the prophecy that "you will earn your living by the sweat of your brow" (Ge 3:19). We consume it effortlessly and without paying anything and in addition it is directly linked to our blood. For this we are equipped with a breathing apparatus which filters oxygen and simultaneously brings it back to our blood. Even though it's a puzzle, can nature happen to do it? Why can't we thank God who reasoned about this, before our creation on earth?

This question brings up another issue, because for science, it is physically through algae that the earth owes its oxygen. So it is, we just tell them it was in God's will, for the purpose of everything he did was to prepare the environment for man. So when the air becomes polluted, the whole organism also becomes polluted and that does not please God according to the slogan "a holy spirit in a healthy body".

However, we are currently inhaling untreated air and making it difficult for our lungs, which is a fantastic organ, which must filter it to pass oxygen to the blood and which the latter carries it through the arteries and veins in all the body. Other gases, naturally occurring in the air, such as nitrogen, are easily released from the lungs because they are designed for them.

But when humans inhale gases that are foreign to the composition of the air, even the lungs no longer know how to control them. They remain attached to it and if they are not toxic in the middle, in the long run, they will prevent the lungs from performing their vital function.

The effects of CORONA VIRUS on the lungs should be of concern to us. Imagine if one day heavy materials filled the air, it would be general suffocation. We are the master of our Universe and by protecting it we are protecting ourselves.

What we will need to know is that no mechanism can truly replace a natural organ with perfection when it becomes defective. Every natural organ has filtrations pores to reject what is foreign to it and filter what is necessary for the body. The failure of rejection of strange bodies by a bodily organ often results in disease.

Rather than playing the empty chair policy in environmental organizations, it's better to say what you don't approve of in their resolution and what you would like to do. This is the case, for example, with the greenhouse gas emissions trading system, called carbon quotas, which does not benefit those who are directly exposed to pollution.

Let's read these articles that demonstrate the devastation caused by air pollution :

– *"Some 7 million people in the world lose their lives each year due to diseases favored by pollution of outdoor and indoor air, according to the latest models from the World Health Organization (WHO)".* (Cf : Anne-Laure Lebrun, Figaro.fr, published on 02/05/2018).

– *"Each year, polluted air causes the premature death of 5.5 million people in the world ... It is six times more than malaria and four times more than AIDS, notes the institution (the World Bank) "*.(Cf : Cécile Thibert, Figaro.fr, published on 09/08/2016).

In addition, in children, fine carbon particles affect not only their respiratory functions, but also the heart and brains.

It is up to us to ensure the protection of the air we breathe and this through the agreements of our various States. No state can escape. Have we thought about the pollution effects of Hiroshima, Chernobyl or Fukushima ?

The important thing to remember is that before any pollution hits the ozone layer or we talk about the greenhouse effect, this gas first passes through our nostrils and leaves its imprint in our lungs. He doesn't need a passport or fill out a business card to do so. We don't even have time to filter it along the way; He does not respect anyone, whether you are a tenant in the White House, the Champ Elyse or the Kremlin Air is the only food we all eat at the same time, as is water for fish. Spills of toxic products have effects on surrounding fish, and even on sea birds. When a vehicle sucks air into the engine and expels it through the exhaust pipes, it is already poisoning the air that enters our lungs. But we cast our eyes on the greenhouse effect by showing only the climatic effects.

We should not analyze things according to the present. To preserve the well-being of the future, we must have a sense of sacrifice, even if our States think that their budgets will suffer or that this will increase unemployment. Should we fight unemployment while killing ourselves ? Who will stay to work ?

The effects of CORONA VIRUS on the lungs should be of concern. Imagine if one-day heavy material filled the air, it would be general asphyxia. We are the masters of our Universe and by protecting it, we protect ourselves. And, the different waves of COVID 19; started since October 2020 would be due to the contamination of the air. When in a country, several people emit the COVID 19 viruses by the blast ; it manifests itself as virtual particles.

At the very beginning of the Universe, when it was still formless and empty as the Bible says, the accumulation of manifestations of virtual particles led to its persistence until they became real particles. This is to say that the more people emit viruses, the more they begin to become resistant in the air and the more the number of infections increases. We are now talking about variants. Just by inhaling the air and not having enough antibodies, it becomes pathology.

In fact, President Donald Trump of the USA, Emmanuel Macron of France and Prime Minister Boris Johnson of the United Kingdom had to suffer from corona virus because of the air infected with these microbes. So when the air becomes impure or infected, nobody is safe.

Nowadays, it is no longer enough to just attack the bodies of individuals, but the cause, i.e. decontamination of the air, followed by decontamination of the air, provided that the spraying is not harmful to the organism.

Noah didn't wait until the day before to prepare for the hurricane. He had taken precautions 100 years before, and that was because he knew he could not prevent this natural disaster from happening.

Ge 5:32 Noah, five hundred years old, begot Shem, Ham and Japheth.

Ge 6:12 God looked at the earth, and, behold, it was corrupt; for all flesh had corrupted its way on the earth.13 So God said to Noah, the end of all flesh is settled before me; for they have filled the earth with violence; behold, I will destroy them with the earth. 14 Make yourself an ark of Gopher wood; you will arrange this arch in cells, and you will cover it with pitch inside and out.

Ge 7 : 6 Noah was six hundred years old when the flood of water hit the earth

We, on the other hand, despite the cry of alarm from environmental NGOs, continue to release greenhouse gases, which also pollute our atmosphere. We are contributing to the smothering of the future and to global warming as in the atmosphere of the planet Venus which, with a strong greenhouse effect, retaining solar energy and increasing the temperature at the surface to nearly

450 ℃, making it the hottest in the solar system and uninhabitable, even though there is air. Even lead will liquefy. But, before we get there, we will already be dead

One of the points which further proves that creation has a master and who ordered everything is the mission which has been assigned to Humankind. Indeed, in the verses below, Humankind is placed above all that in the Universe and is required to take care of it.

Gen 2 : 19The LORD God formed all the beasts of the field and all the fowls of the sky out of the earth, and brought them unto man, to see what he would call them, and that every living creature might bear the name. that man would give him. 20 And the man gave names to all the cattle, and to the fowls of the air, and to all the beasts of the field; "

In addition, humans have the ability to defeat anything that can threaten human life including those caused by any virus or bacteria, even cancer by regenerating tissue roots.

Chapter IX : Cosmology According to the Ancient Oriental Model

Section 1 :
The fundamentals of the two principles (energies)

1.1. Yin and Yang

According to Eastern philosophy, especially Chinese, as developed in the book of transformations, we read this :

"God is the great first origin (Tai ki) of the Universe or origin of the origins. In Him, there are or were released the yang and the yin. From the relationship between yang and yin, the five fundamental elements of the Universe were created namely metal, wood, water, fire and earth. From these five agents, all things came to exist"[1].

Yin represents what is linked to femininity or feminine characters. Yang on the other hand represents what is linked to masculinity or masculine characteristics.

My very poor distinction of yin and yang can be extended from the following table :

1. (Presentation of the Divine Principle, ed. Culture et Paix, Paris, p.28).

yin		**yang**	
black (blue)	黑*hēi*	white (red)	白*bái*
Darkhan (darkness)	暗*àn*	bright (luminous)	亮*liàng*
night	黑夜*heīyè*	DAY	白天*báitiān*
Moon	月*yuè*	Sun	日*rì*
winte (ice)	冬*dōng*	summer (fire)	夏*xià*
north (cold)	北*běi*	south (hot)	南*nán*
left	左 'zǔo'	right	右*yoù*
earth	地*dì*	sky	天*tiān*
female	女*nǚ*	male	男*nán*
empty	虛*xū*	full	實*shí*
introversion	內*nèi*	extroversion	外*wài*
passive	被動*bèidòng*	active	主動 *zhǔdòng*
peer	偶*oǔ*	odd	奇*jī*

Source : "Yin and yang." Wikipedia, the free encyclopedia. Sep 4, 2017.

The explanation of these two symbols which represent the two fundamental tenets of Chinese philosophy was developed when the vocabulary was not so diverse and a single word could represent a lot of things in the understanding of its author. Hence, the non-exhaustive table above.

Thus, for a better development and a good understanding according to our current vocabularies, the yang represents a positive and active energy, while the Yin represents a negative and passive energy. In this way, we come to the conclusion that Yin and Yang are creative forces.

To this day, Yin and Yang are the national symbol of South Korea, and together called Taegeuk, signifies the origin of everything in the universe. We understand why this nation, whose original name Choson was prophesied by the Bible (Rev 7 : 2) as the land of the rising sun (calm morning) from which the angel will emerge who will mark the seal of God's elect. Since the end of the current cosmic cycle predicted by the Mayan calendar, the time has come to reveal all of these things clearly.

1.2. Restitution of the quote from the book of Transformation

https://www.jdbn.fr/yin-yang-signification/

God, according to Eastern philosophy, has in Himself the yin and the yang. From their relationships, all five agents at the origin of all things were created. How did it go?

According to the Universal Cosmological Model, on the one hand, we have water as the source material according to the Bible. After the Big bang symbolized by fire, the elements symbolized by wood and metal, were formed. It was after that came the formation of galaxies, symbolized by the earth. Hence, the sentence : "From the relationship between yang and yin, the five fundamental elements of the Universe were created namely metal, wood, water, fire and earth" should rather be articulate as follows : "from the relationship between yang and yin on water, there was emission of fire, followed by the creation of wood and metal resulting in the formation of the earth (galaxies). And there, we understand why it is written "from its five agents, all things came to exist". This ties in with all that is said in the Universal Cosmological Model.

After the Big Bang, we know that the elements that were propelled gathered again until the formation of the first star. And, it is in this first star that there is formation from simple elements to compound elements such as molecules. And we can say, according to the ancient Eastern Models, that it is through this that the chemical elements composed of the non-metals represented by wood and the metals which were designated by metal are formed.

And after the iron was formed, the star exploded by a phenomenon called a supernova, releasing into the Universe the elements for the formation of other stars and the dust for the formation of the planets, including gases.

For the Bible, this transitional stage is between one evening and one morning after the "separation of the waters from the waters" or Big-bang as we keep reading :

Gen 1 : "6 God said, Let there be an expanse between the waters, and let it divide the waters from the waters. 1 : 7 And God made the expanse, and divided the waters which were in the waters. below the expanse from the waters which are above the expanse. And so it was. 1 : 8 God called the expanse heaven. So there was evening, and there was morning : it was the second day. "

This is how the original order as we drew it in the book of the Divine Principle namely metal, wood, water, fire and earth became, water, fire, wood, metal and earth. The disturbance with the original sentence from which we got the quote was that only the earth kept its position. By referring to the author's vision, we see that the order did not matter to him. What mattered was bringing out the sets that formed the basis of the formation of the universe. The explanations which follow, will logically clarify our classification and that this is not a brain game.

Steps	**Original order**	**Restored order**	**Correspondence with science**
1	metal	Water	The elements of the condensed universe or biblical "waters"
2	wood	fire	Big-bang
3	water	Wood	Formation of life molecules (H, He, C, O, etc.)
4	fire	Metal	Formation of ore
5	Earth	Earth	Galaxies

Section 2 :
Explanations of the primitive elements of the ancient oriental model

We are not going to continue into depth on oriental or Chinese philosophy. What interests us is the hidden explanation of these concepts. For millennia where these thoughts have existed, no exegete has been able to detect their substance. This is due to the fact that science was not as developed as it is today.

As we have said above, religious people are content to speak of the final stage of the creative process rather than starting with the first particle or molecule. Moreover, in ancient centuries, the only way to designate something that could not be explained was to use the name of the finality of the visible compound. This is why today's science does not work with ancient wisdom.

Now, with this manual, science, now understanding this language, will now begin to look at the sacred writings. Thus, we will see what is hidden behind the five elements that Eastern philosophy calls the fundamental elements of creation from which everything that exists is derived. These elements are referred to in Eastern philosophy as metal, wood, water, fire and earth.

We have already arranged the order of these elements. These elements, namely water, wood and metal, already constitute complete bodies and cannot be considered as fundamental particles of the Universe. Hence we will have to dissect each of them to arrive at the Ancient Oriental model.

2.1. The water

Gen 1 : 2 "The earth was formless and empty : there was darkness on the face of the deep, and the spirit of God moved over the waters."

As in Ge 1 : 2, Moses began to describe the constituent elements of the Universe before the Big Bang as "the waters", although according to science, in the void there were already the virtual particles. The author of the book of transformations also mentions among the primitive elements "water". The ancient Western Models which we will see in the following points also referred to as "water". Thus, as in the Universal Cosmological Model, we draw as the primary element before the Big-bang mentioned in the Ancient Eastern Models "water" or "the waters" which represented the original particles, the original materials and the original items.

So, we understand why, the material Pre Big-bang, the consistency of which nobody knows, was described as water by the ancient authors. As I said, even our scientists today fail to give it an absolute name for pre-Big-bang or even Big-bang materials. They are only in the rough, because speaking of plasma, it means that material had already reached an advanced level between particles and atoms. So no one can judge the ancient writers or the Bible on the use of the word "waters" as original material. We have developed this point in Volume II which deals with the pre-Big-bang period to fossil radiation.

2.2. Fire

The fourth fundamental element mentioned in the Book of Transformations as primitive matter in The Eastern Model is fire. However, we know that fire is not a chemical element. As I said, the author of the Book of Transformations ignored the ranking for us to judge him. But, in the light of the Universal Cosmological Model, we know that the name used in ancient times, represented the states of material.

Thus in this context, the fire also is to be shifted in the Pre Big-bang which was an energy sphere and radiation. It is thus like "water", We have developed this point in Volume II.

Fire also represents the expression of Big-bang whose temperature varied beyond $10^{15\circ}$ C.

2.3. Wood

The book of Transformations also cites "wood" as one of the elements of the beginning of the Universe. I have already said that the names mentioned in antiquity concerning the universe, represent states. We know that wood is already the result of several processes, which resulted in the formation of organic cells several million years after the Big Bang.

In fact, wood as a primitive material represents the set of gaseous elements and non-metals, elements that result in life. For us laymen, we only have to burn dry wood to see or feel all of these elements evaporate into the atmosphere. What will remain is dust, which is said to all living things, "You are dust, and you will return to dust" (Ge 3:19).

If the law comes from the Bible, we speak of natural law so as not to glorify its author. But, if it is a scientist who says it, the law bears his name to immortalize him.

Designating all nonmetal and gaseous chemical elements with wood, it makes sense as a designation at the time the book of Transformation was written, as the nonmetals in the periodic table constitute group chemical elements essential to life ; what are called organic molecules.

In addition, the term "molecule of life" does not mean the appearance or development of life. The conditions must be required for life to develop. This is how, in citing planet earth, Moses called it already the bearer of life. This is what justifies the creation of vegetation to plants in Ge 1 : 12-13 on the third day of creation, while it materialized according to Ge 2 : 5-7 on the fifth day which corresponded to the fourth era according to science.

Indeed, according to science, the first elements, called "waters" by Moses, by merging after the Big Bang created hydrogen. The hydrogen atoms colliding, created helium and helium, we ended up with carbon, an essential element for the formation of life, and that in the second or third generation of stars. And about carbon, we read this :

"The whole of life on earth is built around a chemical element : carbon

... Thanks to radiation spectra obtained in different places in the Universe, we now know that the chemistry of carbon is omnipresent."

(Cf : The challenges of science : the origin of life, edited by RBA coleccionables, S.A.U. 2018, Barcelona, Pp.18-68).

The study of carbon compounds gave birth to organic chemistry. According to science, the carbon-14, which is used today in the destruction of fossils, is continuously produced in the atmosphere by cosmic rays and is contained in all living material. So we can find its presence in other planets, meteorites and galaxies.

The designation of wood as all gaseous elements and non- metals alludes to carbon much more in the Ancient Eastern Model.

2.4. Metal

Speaking of "metal" as one of the foundational elements of the beginning of the Universe, the author of the book of Transformations does not see metal in its solid form which is already a very complex process and results millions of years later. the Big-bang. Here? Rather, it is a state of material.

When we talk about metal as primitive material, we see metals in the periodic table of elements. And, here we want to talk about the last element which is formed in the stars and which is often found in the cores of the planets. It's about iron.

In the planets, we speak of iron as being the dominant constituent of their nucleus. Even in gas giants, whose mystery cannot be understood at the moment, we suspect the presence of iron. In addition, it is the 4th element in the earth's crust and it is the most abundant of the metals.

In mining most metals, you always have to get rid of scrap iron. Iron is also found in organic material in small amounts, especially in natural waters, plants and blood. It is a metal that is widely exist in certain chemical compounds with which it associates easily. That is why it is cited as a constituent element of the Universe.

Why did we allude to iron when talking about metal? Because before a star enters the explosive process it produces iron, which, absorbing its energy, signs its death warrant. This Chinese philosophy dates back to around 12 centuries BC.

2.5. Earth

One of the last things mentioned in the transformation book is earth. We said that between fire and earth there was first an explosion or an expansion and this idea is in accordance with the Bible and science. But what does the earth represent here ?

Even Moses after the Big Bang spoke directly of the earth that emerged from the water to which he referred in the following verses :

Ge 1.2 ... and the spirit of God moved above the waters.

Ge 1 : 6 God said, Let there be an expanse between the waters, and let it *divide the waters from the waters.*

Ge 1 : 9 God said, Let the waters which are below heaven be gathered together in one place, and let it be dry. And so it was. 10 God called the dry land Earth, and he called the cluster of the waters Sea. God saw that it was good;

Even though Moses did not take dating into account, the Bible stands to this day as the quintessential cosmological book. In God our Father there is only truth. Scientists know that after the Big Bang, it was very early to be able to talk about the earth. They are right, but to say that we could not talk about the land, then they are wrong.

Indeed, during the period when the elements were projected for the formation of the stars, God was first concerned with what will happen to the earth. He began by positioning the other stars in the solar system, what scholars call the collision period, in order to create a stable place for it to be the bearer of life. It's a phrase that cannot appeal to scholars, but it's like that. This is how the word earth appears directly after the Big Bang in the Bible and the Book of Transformations.

I said at the beginning of this point that what happened after the Big Bang, until the formation of the galaxies, is contained in the biblical period between "one evening and one morning", that is, the evolutionary time to achieve what God wanted.

Moreover, my position agrees with that of scholars, notably that of Brandon Carter's "anthropic principle".

Chapter X : Cosmology according to the Ancient Western Model

As in the ancient eastern model, we do not want to speak of the ancient western model in its historical aspect, but, by deciphering the thought of their authors as we did for the Bible. To this we have retained two philosophers who have argued a lot about the origins of the universe. It is about Empedocles who left us a sandal as a vestige, the other philosopher who is classified as the greatest is Aristotle.

Section 1 : EMPEDOCLE

(Around 490 to 430 BC)

As for the Bible, I am also the only one to be able to discover that the Big bang was also known in antiquity under the name of "Discord" by a Greek philosopher less known in the world responding to the name of Empedocles. Hence we start with this quote that honors him :

"The most colorful figure of ancient philosophy" according to Nietzche, left as a vestige ... a sandal. (Le Monde.fr | 05.19.2014 By Sophie Chassat).

So, just like Nietzsche Friedrich (1844-1900), a German philosopher, who formulated a radical critique of Western thought and Christian morality, and who called Empedocles the greatest ancient philosopher, I too came to the same conclusion.

Indeed, according to legend, EMPÉDOCLES, after having thrown himself into the abyss of the Etna volcano in Italy, which is not probable, because the heat reigning in the surroundings will kill you before even arriving there, his sandal has been either ejected, or the lava did not burn it. The message conveyed by his famous sandal is to show that we will have to follow what he said about how the Universe works, in literary terms of his day and that fits with the Universal Cosmological Model.

Like Moses, Empedocles also ascended a mountain from which he never returned. The body of Moses was even an object of struggle between Heaven and Earth. Read :

Jd 1 : 9 "Now, the Archangel Michael, when he contested with the devil and contested with him the body of Moses, did not dare to bring an abusive judgment against him, but he said : The Lord repress thee! "

1.1. Presentation of the cosmological ideas of Empedocles

Empedocles was a Greek philosopher, poet, engineer and physician from Sicily from the 5th century BC. J.C. He was among the first philosophers who attempted to discover the origin of the cosmos, its foundation and explain how it worked. Even though he is less well known than other philosophers such as Socrates, Pythagoras, Aristotle or Ptolemy, in cosmology I consider him to be the greatest. So my statement may be spontaneous, but Friedrich Nietzsche (1844-1900), who devoted time to the study of all the great ancient philosophers, had come to the same conclusion.

Empedocles attached its name to a theory which, until the creation of modern chemistry, was adopted by almost all philosophers, physicians and esoteric religions : that of the four elements. By developing these concepts on the Universe according to current science as I will unveil it in this book, we will notice that even our scientists are not worth it philosophically speaking.

Let's first read the presentation of EMPÉDOCLE ideas :

> His physical doctrine makes the four elements (Fire, Air, Earth, Water) theconstituent principles of all things. To these Elements are added the Forces ofLove and Hate : Love even brings together what is dissimilar, and Hateseparates what is joined : ...
> At one point one formed from the Many, at another time it split, and fromone came the Many - Fire, Water and Earth and the mighty height of Air.

> Originally, all the elements, united by Love, formed a homogeneous masscalled the spheri. (Hatred sets in) Discord permeates it and separates theelements from each other. Then, after this period of dissolution, Love re- entersthe world and picks up the dissociated elements from Discord. It occurs first ata point (either by the action of Love, or following a rupture of the balancebetween fire and air, so that the beginning of the world would be the finalresult. disorderly movements imparted to the whole by Discord, a whirlingmovement by virtue of which part of the substances are mixed : Discord ispushed out. separated substances are attracted by the primitive mixture ;Discord is driven out more and more. Thus is formed the world, which mustperish again when, after complete unification, it returns to the primitive state ofthe spherus'.
> (Cf : www.Wikip dia. EMPEDOCLE).

In this very philosophical development, we see that, Empedocles speaks of the origins of the universe by a force called love which united everything by forming the spherus, until its destruction by a Big Crunch or the return to l primitive state of the spherus. This makes his philosophy very interesting for the astrophysicists who are reading me now. These writings date from the 5th century BC.

But, first let's continue with his philosophy of how the universe works.

– *"The duality and the opposition of the forces of Love-Hate applying on these four Elements also undergoes an alternation : in a state where Love alone reigns and where everything is united (the Sphairos), follows the progressive introduction of Hatred until the complete separation of the Elements, Love reappearing then brings things back to unity and towards a new cycle"* ...

– *"By the action of the sky, the Earth remains quiet by the effect of a vortex that surrounds it".*

– *"For they alternately prevail in the revolution of the circle, and pass through each other, and become large according to the turn which has been assigned to them.*

In addition, it was the philosopher Empedocles who first proposed in the West a correct explanation of solar eclipses. (Cf : www.WIKIPEDIA : Eclipse).

1.1.1. Scientific reproduction of the thoughts of Empedocles

In the first part of this book dealing with the Pre Big-bang, we made the restitution of the energies of the beginning of the Universe according to Empedocles.

1.1.1.1 The original energies according to Empedocles

1.1.1.1.1. Hate (discord) - darkness - dark energy

For Empedocles "Hate is a force of division and destruction that makes things tend towards the multiple". Like love, hate is described as a force or energy acting differently from love.

On a universe scale, scientists understand directly that it is a phenomenon they cannot explain and have named it "dark energy", which is at the root of the expansion of the universe. What role could it have played before the Big Bang? The answer of the Universal Cosmological Model goes hand in hand with that of the Ancient Western Model or Empedocles. Let's read his thought :

"Originally, all the elements, united by Love, formed a homogeneous mass called the spherus. Discord penetrates there and separates the elements from each other".

By saying that at the origin, all the elements were united by a force that he called Love, it already showed an activity before the Big-bang. Therefore, the pre-Big-bang was never an aporia even in ancient time. It is the scholars of contemporary times who wanted us to believe it.

Also by saying that "Discord penetrates it and separates the elements from each other", we begin the process of Big-bang. For science, it was an instability that crept into supergravity. Gravity became dissociated from the four forces, resulting in the Big-bang. But for the ancient Western model, or even the Universal Cosmological Model, it was a force that came to separate the spherus or the biblical waters. This energy could only be dark energy, which was activated by being confined. Science sometimes speaks of the withdrawal of gravity in the super-force, responsible for the expansion of the original mass.

In other words, one could not say the withdrawal of gravity, but what rather made it jumps.

In these thoughts of Empedocles or the ancient Western model, we are faced with two processes. The first is that all the elements before the Big-bang, which could only be the original particles, were united by a force that he called love. These elements, according to the Universal Cosmological Model, have condensed and warmed up to the extreme, generating the first light.

As we said in the Universal Cosmological Model, before the Big-bang, there had been a cosmic deflation that condensed all the original particles. Empedocles calls this force that caused the deflation, love. The Bible spoke of a wind of God. All believers know that the first attribute of God is love.

The Big-bang was not a random phenomenon. Moses in the Bible interspersed this with the phrase : "God said, let there be a firmament between the waters, and let it separate the waters from the waters" (Gen 1:6). There cannot be a separation without strength. So in front of a wind of God or force of Love that united the particles before the Big-bang, we see that after a short time of lull where the heat reached a maximum temperature in the condensed universe, a force already began to make its entry that will bring about the Big-bang.

Empedocles called this separative force hate, caused "discord". The continuity of his ideas even joins the cosmic inflation by saying that the elements that were close to each other began to unite again despite the expansion as we can read :

"The duality and the opposition of the forces of Love-Hate applying to these four Elements undergoes moreover an alternation : to a state where Love alone reigns and where everything is united (the Spherus), follows the progressive introduction of Hate until the complete separation of the Elements, reappearing then brings things back to unity and towards a new cycle ".

Here we are in the phase of expansion of the universe by the force of hatred and at the same time love struggled again to unite the particles formed the first materials of the universe. These two forces, hate or dark energies for the expansion, and love or gravitation for the formation of atoms and the reappearance of light are described in these terms by Empedocles :

"Then, after this period of dissolution, Love penetrates again into the world and takes back to Discord the dissociated elements". It is understood here that Empedocles speaks of the processes of unification of the elements created by gravitation

in the period of recombination, when dark energy was already in action. He goes on to say that "the beginning of the world would be the final result of the disordered movements imparted to the whole by Discord, a swirling movement in which some of the substances are mixed together : Discord is pushed outwards".

These sentences can be summarized as follows according to the Universal Cosmological Model : after the Big-bang, in order to reach the stability that Empedocles describes as the beginning of the world, there was a war between disordered movements caused by expansion, and the described gravitation of swirling movements that created a mixture of substances. This resulted in the formation, fusion and separation of the stars.

Let's go back to the last sentence of the quotation "a whirling motion that is part of the substances which are mixed : discord is pushed outwards". Thus, here in the formed galaxies, discord or dark energy could not penetrate inside to separate it. It could only push it away, causing the expansion of the Universe. Thus, we see that despite the difficulties of language, Empedocles is describing the formation and functioning of the universe.

With the phrase "discord is pushed outside", we understand that this discord or dark energy no longer had a destructive effect where the elements were united by gravity.

This is how this ancient philosopher, hidden under stones and resurrected by an African, helps us to solve many enigmas.

Empedocles did not limit himself there. He spoke first of the formation of stars and galaxies : the whirlwind slowly expands more and more; the separate substances are attracted by the primitive mixture; Discord is more and more driven outwards. This is how the world is formed". Amen! Here it is first of all a question of the formation of the stars, where there is always a struggle between centrifugal and centripetal forces. Classified this philosophy of cosmogony has no meaning after this scientific restitution. It's thought is indeed in line with what astrophysicists have said about the beginning and formation of galaxies.

1.1.2. Earth's rotation, orbital stability and dark material

Introduced in 1933 by the Swiss Astrophysicist Fritz Zwicky, then consolidated in the 1970s by the American astrophysicist Vera Rubin, the existence or the notion of dark material is integrated into the standard model of cosmology - the Λ-CDM model (for cold dark material) - to explain certain observations, such as the rotation curve of galaxies or fluctuations in the cosmic diffuse background. It also offers an explanation for the formation of the great structures of the Universe and the maintenance of their integrity. It is these last points that interest us and that have caught my attention so that I make a whole chapter in Volume II.

We have said that Empedocles described the orbital stability and rotation of the Earth as follows, "by the action of the sky, the Earth remains still by the effect of a vortex that surrounds it." I want to make it clear that by the words "by the action of heaven", Empedocles is not talking about religion. In principle, he felt that he should have an invisible force which is the basis of the stability of the universe in general and the earth in particular.

Understanding that the ancient philosophers speak of states rather than of names taken in the proper sense, through these writings, we see that he discovered two great truths which present-day science has just discovered : rotation and orbital stability [the Earth (galaxy) remains quiet by the effect of a vortex which surrounds it].

In antiquity, whose horizon was limited to the limit of their eyes, the word earth could represent, depending on the phrase it uses, either our galaxy or the Universe. Here, we understand that he is analyzing our galaxy.

According to our first interpretation, if we can take the earth as being the Milky Way, we can even say that the vortex which surrounds it by analyzing the thought of Empedocles, constitutes the dark material.

Indeed, in classical cosmology, Albert Einstein was the first to discover that there should be a force in the empty pocket of the universe. This observation suggests that the vacuum itself contains energy, in the form of a "cosmological constant" (first introduced, then withdrawn by itself). Other scholars after him will come to confirm it.

The rotation of the stars, the notion of the cosmological stability and the suspicions of dark material, which are the basis of galactic stability and which are believed to have been discovered by modern science, have in fact been known since ancient times.

As for the second interpretation, namely "by the action of the sky, the Earth remains quiet by the effect of a vortex that surrounds it", any moving body creates a wind or a whirlwind around it. Thus the rotation of a star constitutes its self-protection. This is seen in the invisible protections of the earth created by its atmosphere. I also said that in this notion of whirlpool of which the Bible said "a wind from God", we find the notion of gravity expressed in the literary terms of the time.

As for the formation of the stars, Empedocles writes : "For they alternately pre-valuate in the revolution of the circle, and pass into each other, and become large according to the turn which has been assigned to them". Fantastic! There he already describes what is happening in the protoplanetary disc.

1.1.3. The Big Crunch or the end of the world according to EMPEDOCLE

Finally, after having spoken of the expansion which he described as discord and of the gravitation which he called "love", which would be at the base of the formation and the separation of galaxies, Empedocles closes by speaking of the process of the end of the world which is scientifically close to the Big-Crunch by the sentence : *"Thus is formed the world, which must perish again when, after complete unification, it will have returned to the primitive state of the spherus*". This means according to scientific language that gravity will eventually gain the force of expansion and that all galaxies will begin to return to their original position by destroying themselves by "complete unification", until they are recovered in the position of spherus. This phenomenon is scientifically known as the Big Crunch.

This brings us to talk about the future of the universe according to science and not according to the Universal Cosmological Model, because we have not yet had time to think about its end, as long as there are still many things to discover, which will give a precise picture for its future.

THE 3 SCENARIOS FOR THE FUTURE OF THE UNIVERSE according to science

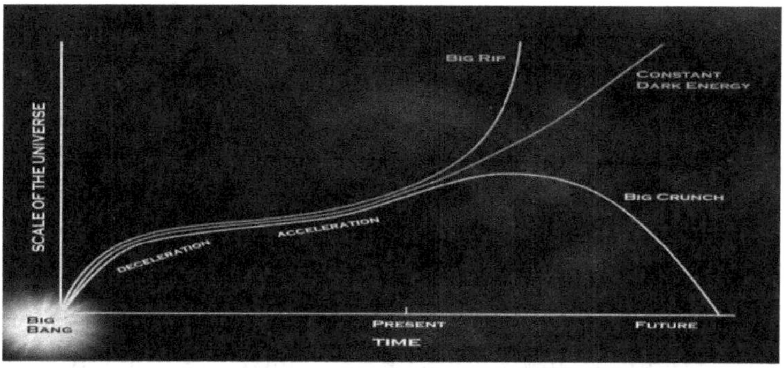

Depending on the variable nature of dark energy or not, the universe will either end up with a Big Crunch or continue to expand forever. In this diagram, we see the deceleration and then the acceleration of the expansion of the observable universe as a function of time on the abscissa. Three possible scenarios for the end of the universe then appear. © Nasa / CXC / M. Weiss.

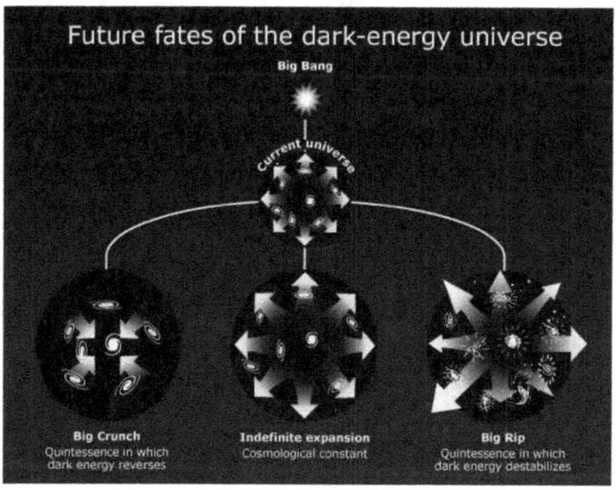

In this diagram, we can see the three possible destinies for the observable universe. © Nasa / STScI / Ann Feild

As shown in the diagram above, the destiny of the observable universe passes through three possible scenarios.

1) It will end with a Big Crunch if dark energy is described by a scalar field whose energy density will vary in such a way that it will eventually become attractive rather than repulsive. It is this scheme that Empedocles foresees for the future of the universe by the sentence "Thus is formed the world, which must perish again when, after complete unification, it will have returned to the primitive state of the spherus ".

This means that if gravity (love) wins, then expansion (discord) will experience a brake and the universe will end up collapsing on itself, i.e. an upside down Big-bang became known as Big-Crunch.

2) If the dark energy which causes the expansion of the universe becomes still so powerful, then even the nuclear forces binding the cores of our body will not be able to oppose a force of repulsion which will dissociate all material structures of the universe. The expansion will manifest itself on smaller and smaller scales, from the level of galaxies to the nuclei of atoms and below. This is the Big Rip or Big Chill scenario, the big freeze.

3) If we are dealing with a cosmological constant, the expansion will last forever and the universe will have no end.

Ecl 1 : 4 one generation is gone, another is coming, and the earth is forever.

Ec 3:14 "I have recognized that all that God does will endure forever, that there is nothing to add to it and nothing to take away from it, and that God acts so that they may fear him."

In this scenario, we are dealing with the "Goldilocks universe", a Goldilocks Universe, where attractive and repulsive forces are in equilibrium without, however, preventing its expansion.

I already want to tell my readers that scientists no longer believe that gravity will be right about the expansion, that is to say the big-crunch, especially as the galaxies are moving further and further apart : "Discord is being driven out more and more" as Empedocles says. But, this scientific scenario always risks ending with a rupture of force and causing the end of the world.

What is the position of the Universal Cosmological Model on the physical end of the Universe ? The Universal Cosmological Model is limited by the scientific restitution of the schema of creation as written in the first two chapters of genesis, and there the option of the end of the world was not addressed. On the other hand, the future of the Universe is sealed by this verse :

Ecl 3:14 "I have known that whatever God does will endure forever, that there is nothing to add to it and nothing to take from it, and that God does so that they may be feared."

It should also be noted that it is this third scenario, announced already in antiquity by King Solomon, which modern cosmology presents as the most probable.

Empedocles' thoughts enter scientific models like that of Fred Hoyle and other astrophysicists and we have called it the ancient Western model. It is up to astrophysicists to continue the development with their mathematical formulas. Like Moses, after the publication of this book, we will also notice that the name Empedocles will no longer be missing in the manual of cosmology like those of Aristotle, Ptolemy, etc.

For those who want to know more about the explanation of the end of the world according to all the prophecies, from the Old Testament to the Apocalypse, the third chapter of the book of divine principle of Rev. Dr. S.M. Moon speaks about it clearly.

We don't want to make comments that will be misunderstood in this book. I put only one of the verses :

Mt 24:29 "Immediately after these days of trouble the sun will be darkened, the moon will no longer give its light, the stars will fall from the sky, and the heavenly bodies will be shaken."

From a geometric point of view, the earth is in the same plane as the stars. Therefore, no star can fall on the other. On the contrary, the earth being smaller than the majority of stars, rather it runs the risk of being gravitational attracted by a star and fallen into it. If this were to happen, for an earthling who thinks the earth is fixed, he will think that it is the star that is falling on him and yet it is the opposite.

These words of Jesus Christ, whose prophecies correspond to the events which will occur when the Sun is going to use up all its hydrogen according to science in about five billion years, are not to be understood literally.

If you are going to tell an astrophysicist that in the last days the stars will fall from the sky (Mt 24:29), even if you are a theology tri-doc, he will think you are a kindergarten student. Bishop St. Augustine said, "If one finds an interpretation of the scriptures that seems to go against what is established (science), this interpretation must be reconsidered." (Cf : documentary "Cosmos - beyond the Big-bang" written by Matt HICKEY). In this way, we will have to review our understanding of this verse. The problem is not the verse, but our understanding. As we have already said, the real explanation of this verse is contained in the book of the Divine Principle. (Cf :

Presentation of the Divine Principle, ed. Culture et Paix, Paris, p.113-122).

1.2 Presentation of the medicinal ideas of Empedocles

For Empedocles, "the combination and properties of the four elements determine health as well as temperaments and characters". This means that the four elements are part of our organism. This sentence is in line with the scientific findings that man was created by the dust of the stars, called dust of the earth by the Bible. What science has failed to say, we have said, is that this dust created life here on earth and not on the stars.

According to the Principle of Creation, branch of the Divine Principle of Rev. Dr. S.S. Moon, to which we align ourselves, this dust is the summation of all the elements of the universe. This is how man is called a micro-universe. Associated with his spirit, he becomes the microcosm of the cosmos. Speaking of the combination of the four fundamental elements that determine health as well as temperaments and characters, it means that these original elements form the individual and that missing one leads to his bio-logical imbalance which manifests itself often by diseases.

And finally, he concludes concerning the evolution of living beings :

"The description of the generation of living beings obeys the same double movement : from a primitive state of androgyny to sexual generation under the progress of Hate ; lonely and wandering members seeking to unite in the reunion phase under the impulse of love. " This sums up his thinking, namely : "The duality and the opposition of the forces of

Love-Hate applying to these four Elements also undergo an alternation".

Referring to the thoughts of Karl Mark, it means that there is struggle and development. Development is being created by a

new unity, advancing history, just as in the struggle of natural selection. These are great notions that I cannot evolve in this book.

Section 2 :
ARISTOTLE

2.1. The primitive elements according to Aristotle

Because of his many works, Aristotle is recognized as the greatest philosopher of antiquity. We even owe him the idea of creating the schools. His thought on cosmology dominated astronomy until the 15th century and this was said to fit in with the doctrine of the Church, of which I do not know on what basis of the verses. For him, the earth was at the center of the universe and all other planets and stars, including the Sun, revolved around it.

According to the scientific documentary "Beyond the Big Bang", the Christian clergy supported Aristotle's ideas only on the basis of an argument namely, "if God created man in his image on earth, the latter can only be at the center of the Universe". (Cf : www.wikipedia, Aristotle).

They were right, except that it was not the earth that was at the center of the universe, but rather Humankind himself. From where we speak of isotropy and this one, it is only the Humankind who notices it

For Aristotle, "The central region of the universe is made up of four elements" as cited by Empedocles : (water, air, earth and fire), and four properties carried by these elements (wet, dry, cold and hot) : it thus constitutes the first physico-chemical model of material. To this must be added a fifth element which is ether. (Cf : WIKIPEDIA : ARISTOTLE).

The object of physics, according to Aristotle, is the study of inanimate beings and their components (earth, fire, water, air, ether). Natural beings, whatever they are (stones, living, etc.), are made up of the first four elements of the Empedocles to which he adds the ether, which occupies what is above the Earth.

In fact, until the end of the 19th century, physicists still believe that space was bathed in a gas, "ether". It was also my conviction. For me, it all depends on what everyone means by what they attribute to that word. Speaking of the ether, is this Aristotle's idea the same as that of astronomers today ? I rather think that Aristotle was alluding to that which cannot be palpated or felt such as the spirituality. As for Albert Einstein, it is more about the physical quality of space.

Indeed, after having demolished the notion of ether in special relativity, Albert Einstein recognized its physical existence in generalized relativity in his study : "Ether and the theory of relativity" where he wrote this :

"To sum up, then, we will say that space is, according to the theory of general relativity, endowed with physical qualities and that in this sense there is ether. According to the theory of general relativity, a space without ether is unthinkable, because in such a space not only the light could not propagate, but also the rules and the clocks could not exist and therefore there would not be of spatio-temporal distances in the sense of physics. But we should not imagine this ether as endowed with the property that characterizes ponderable environments : to be made up of parts that can be followed over time; the notion of movement should not be applied to it. "

And Aristotle did not limit himself to correcting Empedocles there. The latter says Fire / Air / Earth / Water. More logically, Aristotle establishes the series : Fire, Air, Water, Earth. We do not know if Empedocles really took the classification into account for Aristotle to come and correct him. Besides, since these were the constituent states of the universe, I don't know what everyone expected from these sets. As we said, to correct an inventor with the words he used, you have to get into his heart. Perhaps what we call corrections are also what he thought as a whole. The universe is not just words, but sets, that is, structures.

2.2 Meaning of the four sets

Based on the Universal Cosmological Model, I say that fire is a consequence of a collision and cannot be classified as an element. It is a Phenomenon consisting of the release of heat and light produced by the lively combustion of a body. The same is true for air. Indeed, air is the name given to the mixture of gases that make up the Earth's atmosphere. But, nevertheless, both fire and air can carry molecules or particles of their causal elements.

In addition, the fact that these elements are attached to the names of god, demonstrate that they were more about forces, fields and not bodies as we see it. (Zeus, god of celestial light, designates Fire ; Hera, wife of Zeus, designates Air ; Aidoneus [Hades], god of the underworld, designates Earth ; Nestis [Poseidon] designates Water).

For example, at the origin of the Universe, one could not speak of air and fire. Fire was everything that burns, from lightning to the loud bang. This replaces the light and the separation of the waters from the waters of the Bible i.e. the Big Bang. Air represents inflation and expansion. Water represents the elements that are the basis of all creation. The earth is the culmination of the formation of the universe, that is, the galaxies and their contents.

CLASSIFICATION	ELEMENTS	MEANING	EVENTS
1	WATER	Original particles	Quantum mechanics
2	SPHERUS	- Mass Pre Big- bang -Biblical waters	Condensed universe
3	FIRE	- Lights Big-bang	Radiation
4	AIR	-inflation, expansion	Expansion
5	EARTH	Elements	Galaxies

The spherus, representing the condensed universe, emitted the first rays during the Big Bang described as fire. The table above also corrects Aristotle's ranking which corrected Empedocles' ranking. Aristotle said : "Fire, Air, Water, Earth", but this table gives Water / Fire / Air / Earth.

Based on the language of Empedocles, we can say that from the action of love on the waters, the elements condensed to form the spherus. As they warmed up, they finally caused the fire (light) to spring up. After the Big Bang, the name "Air" comes in to demonstrate inflation and the expansion that followed. In the end, galaxies formed represented by the earth. Hence the classification becomes : water, fire, air, earth. I do not know in which manual where Aristotle offered an explanation when he corrected Empedocles, but here I am offering mine based on the Universal Cosmological Model.

Besides, correcting someone who brings something out of the ashes doesn't make you think you're superior to him. In fact, by explaining the behaviors of the energies of the beginning of the universe which he baptized love and hate, from the spherus as a condensed universe, through the formation of galaxies and the expansion of the universe through sentences as "the discord goes further and further away", until it ends in a Big Crunch, Empedocles is part of greater than these correctors.

Regarding the origin of the universe, Aristotle writes : "movement and evolution have no beginning, for the occurrence of change presupposes an earlier process". Here, Aristotle ran into the same problems as our astrophysicists who do not know where to begin the origin of the universe.

Talking about the beginning of the universe supposes an earlier process; hence it is no longer a beginning of the universe. It is this earlier process that becomes a beginning, when it requires yet another earlier process. Hence in Roger Penrose's Model of Cyclic Universes, Moses is the only one who offered a theory of the beginning in the book of Genesis, reproduced in current language in the Universal Cosmological Model contained in this manual.

Aristotle postulates that the universe depends on an eternal movement. Even if he associated God with this eternity, but, his model matches that of Fred Hoyle of a stationary universe. The stationary model theory is no longer accepted by the vast majority of cosmologists, especially after the discovery of background radiation from the sky in 1965 and from quasars.

It was on entering philosophy that Aristotle wrote something that we appreciate in this book namely :

"The entity causing change transmits its form or essence to the affected entity. For example, the shape of a statue is in the soul of the sculptor, before materializing through a tool in the statue".

We retain this sentence to say that we can only see God in man (Ro 1:20) according to this statement of Jesus Christ : "Whoever has seen me has seen God" (Jn 14 : 9).

Chapter XI :
Summary of different models

Section 1 : Synthesis on Ancient Models and scientific models

1.1. The Big bang

According to the ancient Western model introduced by Empedocles, "Originally, all the elements, united by Love, formed a homogeneous mass called the spherus". Science speaks of the condensed universe, while the Bible speaks of the spirit of God or of a wind of God moving on the waters (Pre Big-bang).

Still according to the ancient Western model, "*The duality and opposition of the forces of Love-Hate applying to the spherus also undergoes an alternation : in a state where Love alone reigned and where everything is united, follows 'progressive introduction of Hatred until complete separation of the Elements'*.

For science, the energy that is the basis of the Big Bang still remains a mystery. Some scholars speak of 4 fundamental forces, one of which - namely gravity - has become detached from the others. However, we can see that after the Big Bang, there is still an expansive force which keeps pushing galaxies away from each other, thus ensuring the expansion of the universe. This force is called dark energy. So as in the ancient Western model, gravity and dark energy are part of the duality and opposition of forces that rule the universe.

This duality and the opposition of forces were also recognized in the Bible by Moses from the emergence of light by the phrase "God separated the light from the darkness". To demonstrate that they were now eternal forces, the light was called by Moses "day" and the darkness "night". In the ancient Western model, Empedocles who knew nothing about the Bible spoke of love and hatred for these two dual forces.

Finally, for Empedocles, Discord, which is the energy emanating from the force of hatred, enters the spherus and separates the elements from each other. Science speaks of the Big Bang while the Bible of the "Separation of the Waters from the Waters." So, Moses and Empedocles are the first scholars who mentioned the presence of energy or a force which acted for the Big Bang to take place. Empedocles nicknamed it "Discord."

Moses, on the other hand, introduced it with the phrase "God says" (Ge 1 : 6) to demonstrate the presence of a separated energy.

In the ancient Eastern model, this duality and the opposition of forces are represented by Yin and Yang.

1.2. After the Big-bang

According to Empedocles, in the ancient Western model, "*after this period of dissolution- love re-enters the world and takes back the dissociated elements from Discord. It occurs first at a point (either through the action of Love or as a result of a breakdown in the balance between fire and air)* ". This in parentheses deserves a good explanation.

Scientifically we can say that after the exponential expansion which extended beyond 380,000 years, gravity had new reasons for the expansion and allowed the elements to merge. And Empedocles tells us that it could be the result of either love (gravitation) or a disruption of the balance between fire and air. And we said that the fire represents the set of forces which caused the great heat as the elements condensed until the Big Bang, while the air represented the expansive force.

Out of balance between "fire and air" demonstrates the

deflation that took place before the big bang and the inflation or expansion that followed afterwards. Gravity caused deflation and dark energy or air generating expansion. Thus, to demonstrate that the air of which he speaks is an invisible force and not the element that we breathe, he often uses the expression "the mighty height of the Air" as in the following quote : "At one point, one was formed from the Many, at another time it divided, and from the One came the Multiple - Fire, Water and Earth and the mighty height of Air".

1.3. Formation of the structures of the universe

In the ancient oriental pattern that we have developed in this book, we read this in the book of Transformation :

"From the relationship between yang and yin, the five fundamental elements of the Universe were created namely metal, wood, water, fire and earth. From these five agents, all things came to exist". (Cf : Presentation of the Divine Principle, ed. Culture et Paix, Paris, p.28).

In the ancient Western model, according to Empedocles, - *"the beginning of the world would be the final result of the disordered movements imparted to the whole by Discord, a whirlwind movement by virtue of which part of the substances are mixed".*

Here there is no comment, because the scholars understand very well what Empedocles has just spoken. Indeed, the universe as it is organized today was the final culmination of the disorderly and whirling movements whose bodies broke apart and came together, forming stars and galaxies. So, it is Discord with its expansive energy; what science calls the dark energy that was at the root of these disorders.

He continued. *The tourbillon, at first is very slow, expands more and more ; the separated substances are attracted by the primitive mixture; Discord is increasingly driven outward.* As bodies come together, what science calls "clusters," dark energy is found to be chased out of these set. This is how dark energy pushes galaxies or galactic clusters without causing any disturbance within them. (Discord = darkness or dark energy causing the expansion of the Universe).

And, Empedocles concludes by -*Because they alternately prevail in the revolution of the circle, and pass one in the other, and become large according to the turn which has been assigned to them. Formation of stars in galaxies and each with its dimension according to the bodies captured during the revolution and gravitation.*

1.4. The end of the universe

The end of the universe has been mentioned several times in the Bible and in antiquity by Empedocles. Science has also issued more than the 5 scenarios of which we have explained 3 in this book. But, what the whole model has in common is that we talk about a revival of the Universe in case it happens as expected in the Big Crunch.

In the Bible for example, we read this :

Mt 24:29 Immediately after these days of trouble the sun will be darkened, the moon will no longer give its light, the stars will fall from the sky, and the heavenly bodies will be shaken.

Scientifically speaking, this verse sounds like the phenomena - predicted when the Sun will burn off all of its hydrogen in 5 billion years. These outer layers will swell to such an extent that a living Earthman will think a star is falling on him. But, the explanation of this verse is detailed in the book of Divine Principle of Rev. Dr. Sun Myung Moon in the point concerning the last days.

In the ancient Western model, Empedocles says this : -Thus is formed the world, which must perish again when, after complete unification, it returns to the primitive state of the spherus. This Empedocles' doomsday scenario where everything will revert to the primitive state of the spherus is known scientifically as the Big Crunch. And Empedocles will conclude with a resumption of the universe whose spherus will relax again and that with the phrase "Love reappearing then, brings things back to unity and to a new cycle".

Big-crunch's science scheme also foresees a new Big-bang after the end of the current Universe. In Roger PENROSE's model of cyclical universes, the American scientist develops the same thoughts. Like what, it joins what we said in the introduction to know, in time, the scientists were known under the name of the prophets.

The Bible also speaks repeatedly of the new cycle as we can read it :

Isa 65:17 For I will create new heavens and a new earth;

We will no longer remember the past things, they will never come to mind.

Rev 21 : 1 Then I saw a new heaven and a new earth; for the first heaven and the first earth had passed away, and there was no longer any sea. 2 And I saw the holy city, New Jerusalem, coming down out of heaven from God, prepared as a bride adorned for her husband.

Returning to the ancient Western model by reading the quotes which are very difficult to restore, one understands that Empedocles had a great concern to explain the origins, the composition, the functioning and the end of the Universe. I still want to remember that Empedocles was a philosopher of the 5th century BC and the schema of biblical creation dates from the 15th century BC It is for the first time that these thoughts have been clarified scientifically. This will give birth to other more explanatory books from researchers. From now on, science must recognize its humility in the face of the Bible in general and Moses in particular, as well as this ancient author.

Section 2 : Summary table on the different models

ERA	BIBLE MODEL	ORIENTAL ANTIQUE MODEL	OCCIDENT AL ANTIQUE MODERN	SCIENCE	PRINCIPLE OF CREATION
-5	GOD (LOGOS)	GOD (Great Origin First)	gods		GOD (Origin)
-4	-vacuum chaotic,,			Vacuum (to be included)	Universal PrimaryEnergy)
-3	-darkness - abyss -Spirit or wind of God (Energies)	Yang and yin (Energies)	Love and hate (Energies)	Manifestati on of quantum phenomena (to include)	Universal First Forces (Energies))
-2	- Waters (original materials)	Water	Spherus	Universe condenses	Primary materials
-1	--light (cosmic radiation)	fires	fires	Very hot matter from the condensed universe	Union
0	Separation of water with water	fire	discord	**Big-bang**	(Division)
1	One evening and one morning (evolution)	Wood, Metal	-Air	- inflation darkness, light and fusion of atoms	(union)
2	Earth (Observable universe)	Earth (Observable universe)	Earth (Observable universe)	Observable universe	Observable universe (Multiplication)

Schematically, we can present it as follows :

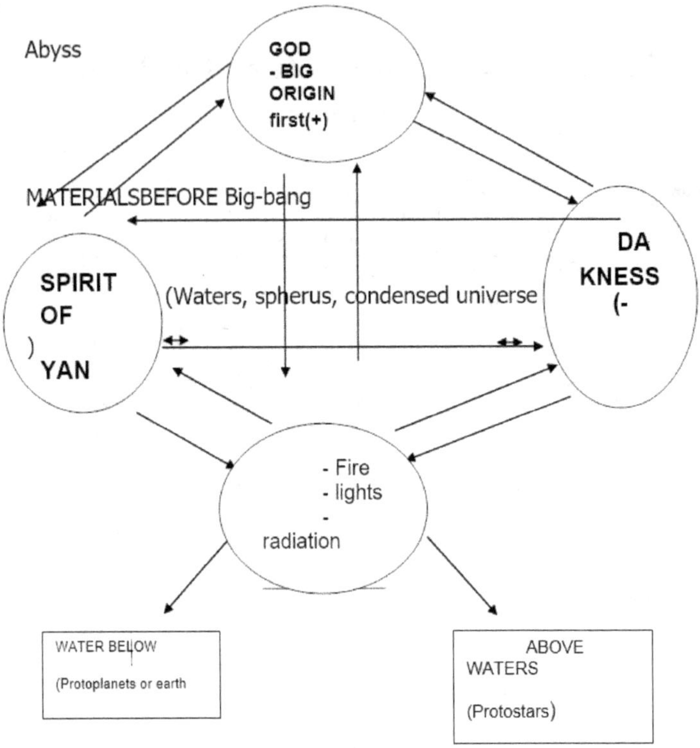

Ge 1 : 6 "GOD separated the waters BELOW with the WATERS ABOVE".

Here we are faced with a transformation of the phases of the original matter called "waters" in the Bible.

As we have just seen, among all the models, including those which are scientifically recognized today, it is the diagram of biblical creation, scientifically restored in the Universal Cosmological Model that is the best. It is the only one, conceived by a single man, Moses, the scholar of scientists, under divine inspiration, who explains, step by step, the evolution of the Universe from the Pre Big-bang, until the Humankind. In our time, to understand how this could have happened, we need the help of many scientific disciplines.

Note that when simulating the collision of the particles to find out what the Universe was like in its early days, scientists to their astonishment discovered that this collision had produced a liquid. And, since they didn't look in the rearview mirror to remember the ancient times when all the great sages spoke of water, they gave their interpretations.

What other name that Moses could use to designate this mass. Let's read what one of the scientists who witnessed the event wrote after the experience and discovery of this liquid. Here she does not speak in the form of philosophical or metaphysical thoughts, but interprets the result after the scientific experiment :

"At the birth of the Universe, for a few millionths of a second, material existed as an extremely hot and dense medium - a sort of primordial 'soup' of particles, composed mainly of fundamental particles of material that we call quarks and gluons". (Corinne Pralavorio : CERN ; From ion collisions at the LHC to record new energy Nov. 27, 2015, 15.41, web article).

In the Large Hadron Collider (LHC) or in any simulation laboratory, we stop the machine to see the result and study it, which already creates a limitation. But, at the scale of the universe, the phenomena were cumulative for millions, if not billions of years. So how much heat did these collisions produce before the Big Bang ? It is a heat that can turn the entire universe into fluid.

Certainly, when scientists speak of the birth of the Universe, they see the aftermath of the Big Bang, because everything is interpreted to conform to the Standard model. But in the Universal Cosmological Model, when we talk about the birth of the universe, our analysis goes ahead of the Big-bang, even though we have accepted the terms Big-bang as being the zero point of the Universe. So the discovery of the liquid after the particles collided was evidence of the use of the word waters in ancient times.

Given the credit that science has given to Aristotle over Empedocles, if he had taken Spherus into consideration, then it would have been known for a long time that the condensed universe and the Big-bang were ancient notions and whose current science has offered evidence. Now all that's left is the scientific approval of the Pre Big-bang as detailed in the One-Verse Cosmological Model in the first part of this book.

Chapter XII
The world of atheists

(http://www.astrosurf.com/luxorion/hawking-hommage5.htm)

The Universe of atheists is part of the vision of the Universe according to non-believing astrophysicists whose arguments, which are not scientifically demonstrable, tend to exclude God as Creator. The various statements of the scientists listed here had only one aim : to exclude God as the creator of the Universe. This is why we could not be indifferent.

When I wrote this chapter, the scientist STEPHEN HAWKING, from whom we have taken many quotations, I wished that he was still alive and would be great if he read it. While waiting to be edited, I learned of his death through the media. I know that the debates between scholars only concern theories and nobody questions the intelligence of a colleague. Some have had to accept their mistakes, but this has not tarnished their images in any way.

What I write here are only my reactions to the unprovable ideas of some scientists on the question of creation.

My work too is not perfect, and I would even say that it is a test to which I take full responsibility for my mistakes. The physicist Niels Bohr said :

"The opposite of a true statement is a false statement. But the opposite of a profound truth may very well be another profound truth." (Cf : J.Baker : 50 keys to understanding astronomy, ed. Dunod, p.70)

In this chapter, we have taken most of the quotations from the scientific documentary "The Universe of STEPHEN HAWKING" where this great British theoretical physicist exposed what he thinks from of Big-bang to man. We point out that Stephen HAWKING's conclusions do not constitute a model, but influence the thinkers of the standard cosmological model.

Moreover, to answer the concerns of the scientist Stephen HAWKING, I could not but enter the Universe of the cause, because his concerns are much more likely to expel God as Creator. I am making a whole chapter of them, because these ideas allow me to respond to persons who share the same ideas as him.

Most of my arguments in this chapter have already been developed in previous pages. The titles of the different sections are the answers reserved for the conclusions of atheists in general and the statements of the scholar Stephen HAWKING in particular. I would also like to make it clear that a great thinker who has left traces never dies, but always speaks through these writings. So the death of a great thinker never closes the debate on what he left behind and it allows him to never be buried.

Section 1 : The purpose of the Universe's existence

1.1. Why was the Universe created ?

"My goal is to understand why the Universe exists" (Stephen HAWKING)

Prof. Stephen HAWKING mentioned that he puts a lot of emphasis on this issue, which in principle, is not part of a model.

To answer this question, one must refer to the origin, and the origin is none other than God the creator. Can a dog say why it was bought ? It will only be important to its owner if it fulfills the purpose it has been given.

Indeed, many people, including believers, ask themselves the same question about the existence of the Universe and man. Only God could answer this.

If I buy or make a knife, only I know why I made or bought it. All those who are outside me will only speculate. Thus, for this answer, only Christ, who as witnesses to the creation of the Universe (Jn 1:2) can answer this and it is through the body that incarnates him (Jn 1:14). As these two verses attest :

Jn 1:2 "The word was in the beginning with God".
Jn 1:14 " The Word became flesh and made his dwelling among us. We have seen his glory, the glory of the one and only Son, who came from the Father, full of grace and truth. ".

Now, throughout six thousand years of religious history, there have only been two persons to whom Christ has been perfectly incarnated. First there was Jesus Christ and 2,000 years later it was the Moon couple as Lord of the Second

Coming. There is no need to sulk because God's choice was never a democratic choice. It does not depend on the one who wants to, nor on the one who runs, but on him alone (Rm 9:16). Is it not also written that he had no beauty or radiance to catch our eyes (Isa 52:2) or that the kingdom of heaven does not come in such a way as to catch our eye (Lk 17:20) and that only those who go to hear him will discover the truth (Rev 2:17)? Thus, it is only to these two sons of God that we can have the answer to the question "why the Universe was created".

As I said at the beginning, Jesus judged the intelligence and level of understanding of the people of his generation to be low (Jn 3:12) and promised that the Paraclete will come and tell everything. Thus, it is in the Divine Principle of Rev. Dr. Sun Myung Moon, living in a generation where the spiritual and intellectual level of men is greatly evolved, that we find the answers in these excerpts :

"The purpose for human existence is not determined by the created being himself; the true purpose is determined by the creator of human. We must therefore know God's purpose for creation in order to be able to understand the true purpose of man and the universe... It is the nature of the heart to seek an object to love. This nature of the heart is the motive for which God made creation. This means that God, whose essence is the heart, feels joy when He can love an object, an object that He created. Without an object, God cannot satisfy His impulse to express His attention and love, which gushes forth like an inexhaustible spring. God made creation to be the object of His love...Joy is felt by the stimulus that a subject feel when his own nature is reflected in an object, whether the object is visible or invisible... Thus, we understand that God feels joy when He feels that His Original Sung Sang and His original Hyung Sang (His inner and outer nature) are reflected in His creation". (The Divine Principle Level 4, P. 26).

So, in this quotation, we retain 3 answers.

- The purpose for which we exist is determined first of all by the creator himself, i.e. God.
- This purpose is achieved when the creature (object) reflects what the creator (subject) expects of him, the most important is to give back to the creator his image or to reflect his image (Mt 5:48).
- Hence, the Universe was created to become the kingdom of heaven.

For this reason, unity between science and religion is necessary, and this through the new branch that we call "astrotheology" ; a branch that complements all disciplines.

1.2. What is man made of?

In all scientific documentaries, we are told that man is the dust of the stars. What does this mean? Let's see the different statements first :

"We are made of the foam of the Big-bang" (Stephen HAWKING)

"We are made of the debris of stellar material that has cooled by accident, the debris of failed stars" (Sir Arthur Eddington).

"We are made of stardust" (astrophysics)

"We are made of the dust of the earth" (Bible)

"We are made of mud" (Qur'an)

"We are the microcosms of the cosmos" (Dr. Sun Myung Moon in The Divine Principle)

All of the above statements are based on scientific thinking and not on faith, even those in the Bible and the Qur'an. In order to answer this question, scientists have referred to the origin of the solar system. According to science, all the planets of the solar system were projected by the explosion of a star, from which part of the dust and gas, having formed the molecular clouds, collapsed gravitationally. 99% of the matters contained in the center formed the Sun. The other scattered matters representing 1% formed the planets of which the Earth.

The formation of a star is made from molecular cloud called also nebula which is composed largely of hydrogen molecule. By the gravitational contraction of a nebula, the chemical and physical processes are triggered. We begin the nucleosynthesis which is the set of nuclear reactions that occur inside the stars. Hydrogen is the major fuel of stars and its fusion is the first link in the chain of nucleosynthesis.

By gravitational collapse, hydrogen starts to contract. Because of this contraction bringing protons closer together and thus increasing their collisions, the temperature gradually increases, which will lead to their fusion, resulting in the formation of helium. The contraction continues its cycle, causing more and more the increase of the temperature and the release of energy, this one ended by the fusion of hydrogen and helium, and by the creation of the various elements in a progressive way such as carbon - nitrogen - oxygen, To not describe all the cycles.

It is necessary to know that at each creation of an element, the mixing with the others lead to the creation of another element. We can see the order of this creation in the table of Mendeleev by starting with hydrogen. These processes continue until the formation of iron. From hydrogen to iron, it can take millions, even billions of years within the star.

Iron being endothermic, that is to say consuming heat, the star begins to absorb energy from the environment instead of emitting it. This will result in the explosion of the star, releasing a strong heat and light called supernova, while violently ejecting into space, under the effect of the shock wave, the heavy elements it has synthesized. This creates more heavy elements outside the star. It is from such an explosion, that part of the dust from this star called stellar nebula that will form the solar system including the earth where living beings will appear.

It is from this upstream analysis that astrophysicists say that man is the scum of the Big-bang, the debris of stellar matter, the dust of stars. But, analyzing the situation downstream, the Bible and the Koran are also right to say that man is the dust of the Earth, because it is on the Earth that he was formed. The Qur'an speaks of mud because without water, life cannot appear. Mud means dust plus water, and man is composed of 70% of water and 30% of the dust of the earth (see the composition of a multivitamin medicine). But, Dr. Sun Myung Moon went even further by taking into consideration the mind because without it, the body is only dust.

Indeed, we have said that the universe and the spiritual world form the cosmos or the macrocosm. Man being the mixture of these two entities, that is to say the spirit representing the spiritual world and the body representing the universe, is thus qualified by Dr Sun Myung Moon in the Principle of Creation as a "microcosm of the cosmos".

For Stephen HAWKING, we are formed from the foam of the Big-bang. Foam is defined in the Grand Larousse Robert as "*a whitish Mousse, with or without impurities, which forms on the surface of agitated, heated or fermented liquids*". I don't know whether he speaks as a Prophet or an astrophysicist. His colleagues say that we are made of stardust and the Koran speaks of mud. As you can see, the whole point of view revolves around the word dust, whether it is foam or debris of stellar material. Moses completed "earth" because after all it was the final destination where man was formed.

The Bible and the Koran presenting the situation downstream, develops the same points of view. Moses spoke of the dust of the earth, i.e. all the material of the Universe as explained in the Principle of Creation by Rev. Dr. Sun Myung Moon. Science added that without water in a liquid state, life cannot appear. When we take the dust from the earth according to the Bible and add water to it, we get mud as predicted by the Qur'an. Isn't it written "God formed man from the dust of the earth? "(Gen 2:7). But before that, Moses had to raise the issue of water (Gen 2:5-6), thus joining the Koran and science. In other words, science confirmed the Koran and the Bible.

Gen 2:5 There was not yet a shrub of the field on the earth, nor any herb of the field sprouting, for the LORD God had not caused it to rain on the earth, and there was no man to till the ground

So, the scientists analyzing the situation upstream, have confirmed with technical terms what these two sacred books have said downstream. But, what is different between the religious concepts that are part of the Universal Cosmological Model and science is that animation can be created in an object from the two opposite poles seeking to unite, but consciousness is not created. All scientists have observed this; material alone, even animated, is not enough to create consciousness.

That is to say, in addition to the Big-bang foam or stellar debris approved by scientists, God had to do something more; he blew into man and man became a living being (Gen 2:7). I had already said that this breath had nothing to do with the air we breathe, which is also part of material or a set of materials. It is the divinity itself that has been transmitted to us, including consciousness.

In the scientific documentary "From the Big-bang to the living"; written by IOLANDE CADRIN ROSSIGNOL, the scientist's last conclusion was : "It is the appearance of consciousness which is the greatest miracle in the Universe". Is there a scientist who can say something about the appearance of consciousness which is immaterial? Only the principle of creation, a branch of the book of the Divine Principle by Rev. Dr. Sun Myung Moon, which explains it in detail.

We can read his answer through this quote :

"If you enter into a mystical state, you will understand that a single grain of sand contains all the principles of the Universe. You will see even in a single atom the infinite and inexhaustible harmony of the Universe. Despite our limited understanding, we cannot deny that a complex network of forces is at the origin of all existence. Molecules contain atoms and atoms contain elementary particles. Yet even these entities are not without consciousness. Each one is endowed with a consciousness and a specific goal. (Cfr : Cheon Seong Gyeong, FFPMU, p.85)

In the past, scientists denied that all the bricks of life were made in the stars. But nowadays, other scientists are beginning to join the Universal Cosmological Model, i.e. man is formed from the dust of the Earth as we can read in the following quote :

"Some biochemists believe that even the appearance of life owes a lot to the tides, which would have favored the creation of amino acids, the building blocks essential to the constitution of the first living organisms, some 4 billion

years ago.".(Cf: Hugo Leroux, Science&VIe Hors-Série n°288, 09/2019, p.58).

Therefore, not only can amino acid synthesis also occur from stellar dust according to science, but it can also be favored by stormy weather conditions on earth

1.3. The motor of life

"What amazes me is that we are made of the same material as the stars. If our heart beats, it is thanks to the energy released during the transmutation of these materials". (Stephen HAWKING)

Believers will react by talking about the grace of God instead of *to energy released during the transmutation of these materials.* But, scientifics do not like such explanations even if it is supported by biblical verses they do not even believe in. Besides, most of the verses I put here I do that for the believers who themselves also have doubts to believe my arguments. The only detailed scientific explanation of the energy whose source comes from God, giving and keeping us alive, is in the book of the Divine Principle by Rev. Sun Myung Moon. A book that resolves and gives us the wisdom to solve all the concerns of contemporary men such as those raised by Karl MARX, Jean-Paul SARTRE, Stephen HAWKING and why not ourselves. I have already spoken about it at the beginning of this book, but let us make a synthesis.

As I said, we can never talk about the origin of a thing in creation by excluding God. If an atheist does not want us to talk about God, the truth wants us to associate God, so we cannot help but talk about God.

All things, to which we allocate an existence, whether spiritual or physical, must have energy in themselves. According to The Principle of Creation, a chapter of Rev. Dr. Sun Myung Moon's Divine Principle, the energy that is in God is called "the Universal Prime Energy". Let us not forget that the name is given by man. So, in God there is the original energy. The one who first discovered this concept, Rev. Dr. Sun Myung Moon, called it "The Universal Prime Energy" ; a name that science should also adopt. It is this energy, the very basis of God's eternity, which is endowed by God to every created entity and is the fundamental force of creation.

The endowment that was made from the first living being, however minute, was illustrated by Moses at the time of man's creation with the following verse :

Gen 2:7 the LORD God formed man from the dust of the earth, breathed into his nostrils the breath of life and man became a living being.

Here we are dealing with an inert entity, which according to the astrophysicists is made of the dust of the stars. And then the quote from Stephen Hawking continues : "If our heart beats, it is thanks to energy". Once again Stephen Hawking discovered a great secret, but he allocated the source of this energy to the transmutation of material. Moses also used the synonym of transmutation by the word blow, but, this breath or energy came from God and it is well explained as I said in the Divine Principle of Rev. Dr. Sun Myung Moon.

So, the answers to Stephen HAWKING's concerns will enable many of the scholars, and there are not even many, to believe in the existence of God. Believing in the existence of God has nothing to do with believing in God which means worshipping Him. Even Satan believes in the existence of God. It is as if I could say : Mr. Macron is the President of the Republic, but, as a Congolese, he is not directly my President. However, this does not exclude that I pay him the honors linked to his rank.

Section 2 :
God and Coincidence

2.1. Are we intelligent monkeys ?

"*Without a new cosmic coincidence, there would be no man on earth. Without imperfection, neither you nor I would have existed. We are only advanced intelligent monkeys*" (Stephen HAWKING)

All the world's religions, taken together, attribute to God the image of a perfect being and the creator of the Universe. For Christians, man is the son of God and is not to be compared with an animal in the true sense of the word, although biology does not distinguish between them.

For Stephen HAWKING, the Universe did not need God to be created. It is only the result of physical processes. And to further dismiss God, he concludes that the present Universe and everything contained within it, down to the men who are only monkeys with advanced intelligence, are the results of imperfection and chance.

Listening to this statement, I wondered whether coincidence could produce two complementary beings like a male and a female for a perpetuation of the species at all levels. Scientists and other atheists do not like such examples. There is already an abundant literature on this, but it hasn't changed any of them. And, they don't even read literature that tends to prove that God created the Universe. And, they don't even read literature that tends to prove that God created the Universe. The only proof for them is the scientific proof that the Universe was created by God. Moreover, also scientifically proving to them the existence of God. This is what we have done in the Universal Cosmological Model by restoring with the scientific literature, the history of creation according to the Bible.

For the second question, I recommend to them the analyses of Rev. Dr. Sun Myung Moon contained in the Principles of Creation, which is one of the teaching chapters of the book of Divine Principle.

For the believers who are fooled by phrases such as man is an intelligent animal or a monkey with advanced intelligence, I tell them that all animals have advanced intelligence. It is rather their culture that is different from ours. What's more, man is the divinity, even if we are still far from our identity.

At the time of creation, God made a clear distinction between animals and man. From Genesis 1:20-25, God spoke of the creation of animals, but as soon as man was created, the language took a new turn. God recreated himself :

Ge 1:26-28 "Then God said, "Let us make man in our image, according to our likeness. They will rule the fish of the sea, the birds of the sky, the livestock, the whole earth, and the creatures that crawl[g] on the earth.27 So God created man in his own image; he created him in the image of God; he created them male and female.28 God blessed them, and God said to them, "Be fruitful, multiply, fill the earth, and subdue it. Rule the fish of the sea, the birds of the sky, and every creature that crawls on the earth.

Gen 1:31 And God saw all that he had made and it was very good indeed. And there was evening and there was morning, the sixth day.

Here we see all the care God put into man's creation. It was followed by the expression "very good". God did not say : let us create an intelligent animal that resembles us, but "man in our image, in the image of God". It is this expression that God wants each one of us to look at our neighbor and consider him or her in this way, regardless of race or social rank. This is what will put an end to the quarrels, which on a national and world scale lead to wars. Stephen Hawking himself understood in the end that man is a being apart from the animals through this statement :

"*We are only an advanced race of monkeys on a minor planet revolving around a very average star (the Sun). But we are capable of understanding the Universe, which makes us something quite apart.*"(For : J.Baker : 50 Keys to Understanding Astronomy, ed. Dunod. Paris, 2016, p.14).

2.2. Does the Universe come from nothing?

"*Everything is born from nothing*" (Stephen HAWKING)

Ge 1:2 "The earth was formless and empty", (Louis Segond version)

Gen 1:2 "The earth was invisible and empty, and there was darkness over the surface of the deep, and the Spirit of God was moving over the waters" (Septuagint Version)

Here the scholar Stephen HAWKING is right, because the Bible also confirms this with the sentence "The earth was formless and empty". I had already explained that something that is formless and empty simply means something that exists, but has no form and cannot be seen. It is simply the Universe at its beginning where everything was nothing, that is to say an endless vacuum but energetic..

However, when the first Pre-Big-bang original particles began to agglomerate under the action of the "Spirit of God" or "a wind of God", scientifically restored by us as the First Universal Force, there was the first deformation of the fabric of space-time, described by the Bible as "the abyss", and no scientist can contradict this, because there is no basis for denying this restitution of the Pre-Big-bang Universal Cosmological Model.

Using the phrase "the Universe was born out of nothingness", compared to other scholars who advocate the stationary model, Stephen HAWKING nonetheless acknowledges that the Universe had a beginning as mentioned in the Bible, although its nothingness excludes God.

2.3. Did God create the Universe?

"*The Universe sprang from itself in the Big-bang*" (Stephen Hawking)

"*Because of the law of gravity, the Universe can create itself out of nothing*," Hawking wrote in his book The Great Design. (2014)

I have already raised the fact that scholars have made the Big-bang a singularity, that is to say, where it all began. But, some, like Hubert REEVES in the scientific documentary " Big-bang alive ", have made it a horizon, that means, even if they don't know what happened before, but at least they still suspect certain phenomena. Many of the believing scientists classify themselves according to this point of view.

From the time of the publication of the series of books "THE WAYS OF SAVINGS", scholars should know that the Universe was not born out of the Big-bang and that the events before the Big-bang are scientifically well explained and exposed, that is more than 1,500 years BC by the scholar of scholars Moses in the Bible.

In fact, taking the Jerusalem version of the Bible we read the following about the Pre-Big-bang or pre "separation of the waters from the waters" according to the Bible :

Ge 1 : 2-4 "The earth was formless and empty : there was darkness over the surface of the deep, and a wind of God moved over the waters, and God said, 'Let there be light'. And there was light. And God separated the light from the darkness. That was the first day.

In Volume II, which is entitled "Big-bang to fossil radiation», I had tried to explain the great ensembles, in particular :

– Darkness : the seat of dark energy, which is still an enigma for science even though it has already confirmed its existence;

– The abyss : a distortion of the fabric of space-time, caused by the biblical "waters" or primordial particles from the Pre-Big Bang;

– A wind of God : the movement that manifests itself at the contour of the abyss (black hole) creating gravitational waves, which some scientists have even won the Nobel Prize for discovering in 2017. But Moses spoke about it 1500 years ago before Christ. On the other hand, it was Albert Einstein who issued only the principle that led to its discovery. After the discovery of gravitational waves, Mr Grichka Bogdanov, author of the book" la fin du hazard" or 'The end of coincidence" (Grasset, 2013) gave his analysis to express which we can read :

"These gravitational waves could be, in our opinion, the trace of phenomena whose source is perhaps "before" the Big Bang. In other words, when physicists rightly indicate that the waves in question express "fluctuations" in space-time, in reality the origin of these fluctuations could well be to be sought even before the Big-bang, i.e. before the explosive birth

of material, space and time. (Cf : Express.fr, the first tremors of the Big-bang explained by Grichka Bogdanov, by Jérémie Pham-Lê, published on 18/03/2014).

– Water : the first quantum primordial particles that later led to the creation of hydrogen, which according to Lavoisier means "that which generates water". This effectively fits with the name hydro in Hebrew that Moses used. (Cfr : WATER, wikipedia).

– The light was : this was physically the first consequence of the whirlwind described in the Bible as "God's wind was hovering over the waters". We have said that it may have been the fossil radiation that two American scientists, Arno Penzias and Robert Wilson, received the Nobel Prize in 1978 after also discovering it. It was not a prediction by George Gamow but by Moses, who, by calling it "Day", described its eternal presence and whose WMAP satellite photographed it 1413,8 billion years after the Big Bang.

Until then we have described the appearance of this biblical light of fossil radiation as a prediction. This is because science places the appearance of the latter at 380,000 years after the Big Bang. But the majority of scientists are already beginning to confirm that this light did not appear, but rather reappeared. But when did it first appear, since at the extreme condensation of the Universe before the Big-bang, science says that it was no longer able to emit light ?

This new enigma is followed by another enigma : before the Big-bang, light and darkness being energies, why did Moses say that "the two were separated" (Gen 1:4) ?

Therefore, Moses is still ahead of current science and this is no reason for our scientists to ignore the period before the Big Bang, which was very lively. This book has just planted a new way of thinking in science, so that in the centuries to come, these phenomena will be scientifically confirmed.

One thing is certain, we have said; after the publication of this book, no scientist will say any more that the Universe was born from the Big-bang. The Big-bang now becomes a horizon. With the consideration of the scheme of the Universal Cosmological Model, scholars will now be breaking their heads to understand the scheme of biblical creation as well as theologians to understand the biblical parables and the thoughts of the authors of sacred books instead of relying only on the etymologies of words.

Moses could not alone in his time issue such a schema with much precision, but must have taken it under the inspiration of a Being who was either a witness or the creator. And there we come to God. But this in no way detracts from the intelligence of Moses, who knew how to translate what comes from God into everyday language and which today science is confirming. Hence, this leads to his elevation to the rank of Scientist of Scientists.

Section 3 : Big-bang : separation, expansion or explosion?

3.1. The Big-bang comes out of the darkness

"*The Big-bang occurred in complete darkness because there was no light*" (Stephen HAWKING).

The question of whether the Big-bang occurred in complete darkness is not a concern in the Universal Cosmological Model. Initially Father Georges LEMAITRE spoke of the explosion of an primitive atom. and this can only generate the boom and light. This is what prompted the scientist Fred Hoyle to describe this phenomenon as a "Big-bang" in the form of mockery. And, it was always in this idea of an explosion that Georges Gamow addressed that there should be residual traces, and it was discovered and science called it fossil radiation.

In addition, the Bible refers to "waters" as material from before Big-bang. Father Georges LEMAITRE spoke of a "primitive atom". The debate on its dimension is not yet closed, as Dr Harold S. SLUSHER stated in the scientific documentary "The Origins of the Universe" by prof. Dr. A.E. WILDER Smith.

After the observation by the American astronomer Edwin Hubble in 1929 that galaxies were moving away from each other, Father Georges LEMAITRE spoke of the expansion of the Universe and this is what science now adopts, just as science has adopted that for 380,000 years after the Big-bang, there was deep darkness.

Moses through the Bible in the creation story contained in Ge1 : 1-6, having spoken of the emergence of light before the Big-bang, also did not use the term explosion followed by light after the Big-bang or "the separation of the waters from the waters".

To demonstrate that the Big-bang was part of God's will, Moses used the word "separation of the waters from the waters". The waters represent what science calls the condensed Universe before the Big-bang. If a road splits in two under the effect of torrents, we speak of breaks. But if this was part of the government's will, we talk about "separation". But, after the separation, Moses speaks of the expanse that shows the formation of a vacuum. Moses spoke of the phenomenon in the present tense, but the dynamics of the stars in the vacuum can only create expansions.

Moreover, a separation or swelling of an atom, followed by an expansion that after a thousandth of a second already reaches the size of a balloon, does it not deserve the name of explosion ?

Section 4 : Universal Primary Energy is the driving force behind everything

4.1. Energy and gravity

"*The driving force behind everything is gravity*" (Stephen HAWKING)

To respond to someone, you first have to find out if you have understood their idea. The Law of Universal Gravitation indicates that two objects with mass exert a defined force of gravitational attraction on each other.

Thus, our answer is based on the case that for Stephen Hawking, gravity can be considered as the fundamental force of everything. If this is the case, then this is not true for the Universal Cosmological Model, since gravity is the resultant of different phenomena that can only be observed in the macroscopic world. Moreover, it is only one of the four fundamental forces of interactions of material and is not even included in the standard model. The other forces are the strong and weak nuclear interaction forces and the electromagnetic interaction force.

Prof. Peter Ware Higgs, a British physicist, said that particles only bind together after they have acquired mass and this through a boson called Higgs' Boson. In order to acquire this mass which will give them the gravitational properties, according to current physics, we believe that they must first possessed energy.

In fact, the atomic nucleus is made up of a proton and a neutron. The proton is a hadron composed of two up quarks and a down quark. The neutron is made up of two down quarks and one up quark. This differentiation, noted as energy (++-) and (– -+) in the Universal Cosmological Model, is the scientific basis of all waltzes in the Universe. The nucleus forms an atom by associating itself with the electron. The electron that revolves around an atomic nucleus is equivalent to an electric current; it thus creates, perpendicular to its orbit, a magnetic field. More generally, magnetism, and in particular the magnetization of material, is the consequence of arrangements of the elements (spins) of the electrons. (Cf : Encarta; Electron).

We should know that gravity is a consequence of the dual characteristics of God endowed to creation as we can read in these sentences :

"*The Universe is the unfolding and manifesting of the inner nature and the original outer form of God according to mathematical principles. It is therefore clear that there is a mathematical aspect to the nature of God*". (Cf : Présentation du Principe Divin, ed. Culture et paix, Paris, Pp 26-43).

Thus, what are these mathematical principles which contribute to gravity and which are part of the dual characteristics of God? The Principles of Creation, the chapter on the Divine Principle by Rev. Dr. Sun Myung Moon explains them in detail, but we will stick to the effects that science has already discovered. Here we quote Prof. Higgs' conclusion, Newton's law, Coulomb's law and Einstein's relativity.

– According to Prof. P. Higgs, as we have said, "the fundamental particles only manage to unite after having acquired mass... (Cf : documentary "The Secrets of the Cosmos : Higgs' Field Theory"). So here we have in the particle, the relation energy-mass, that is to say, internal character and external form.

– The second phenomenon derives from Coulomb's law and Newton's law. In fact, it is scientifically proven that Coulomb's law is Newton's law applied to electromagnetism. Isaac Newton's Law of Universal Gravitation indicates that two objects with mass exert on each other "an attractive force" due to the force of gravity released by each of them. Coulomb's Law, on the other hand, concerns the magnetic force of two bodies. It says that two charges of opposite sign attract each other, while charges of the same polarity repel each other. We thus speak of charges or positive and negative poles within particles and in the mass or magnetic force. Several formulas have been emitted putting these two laws in interaction. Putting it in this book and explaining it will put my readers to sleep.

– Then Albert Einstein's famous discovery comes, which says that mass is concentrated energy.

So energy-mass, positive – negative is part of the dual characteristics of God who has energy as an attribute. The Particles are also made up of energies, each of which carries a dominant pole of one charge, the other being latent. It is through this that the myriad relationships in the Universe arise. But here we want to see how gravity is created and wise scientists already have the answer on the basis of mathematical principles put together in a formula. Today's gravitational forces are influenced by the cluster of belonging bodies.

To return to the way gravity is formed, we use the Principle of Creation, the chapter of Rev. Dr. Sun Myung Moon's Divine Principle, so that scholars and theologians know that God's word evolves with time and no verse can block God from speaking on the concerns of the day, be it on the revelation of all the biblical parables, providential history, theology, politics, economics and science. In this book I exploit only a tiny part of the first chapter of Rev. Dr. Sun Myung Moon's Divine Principle which speaks of

God and the Universe in the branch called the Principle of Creation.

I was saying that the electron and proton contain the dual characteristics of God as described above :

"*When an electron forms a common base with a proton and engages with it in an action of giving and receiving, it revolves around it in a (circular) movement. Thus they unite and form an atom (hydrogen). The electron and the proton are themselves made up of dual characteristics which are at the origin of their continuous rotation. This is why the circular movement born from the action of giving and receiving between the proton and the electron is not limited to an orbit in a single plane but, by constantly modifying its angle of revolution, creates a spherical movement. Because of this spherical movement, the atom exists in three dimensions. In the same way, the magnetic force between the positive and negative poles causes electrically charged particles to describe a spherical movement*". (Cfr : Presentation of the Divine Principle ; ed. Culture et Paix, Paris, P. 36).

My biggest surprise, I who never did theoretical physics, was the fact of putting in theoretical relation the laws of Higgs, Newton and Coulomb to explain the origin of gravity. I was surprised to find that science had already done it and had even already issued formulas called the COULOMB-NEWTON LAW in magnetism (weak force) where Higgs boson-vehicles are integrated, that is to say, they are the pilots of the magnetic inducing field towards charged particles (leptons, baryons, mesons).

But, in order to bring out the driving force of everything, which would not have been the concern of astrophysicists and myself if Stephen Hawking had not spoken about it, I said that I was still introducing Albert Einstein's law of $E = mc^2$, (based on restricted relativity and material-energy equivalence) without referring to his concern. The surprise still followed when I noticed that already in 1916 Einstein had published the theory of general relativity, which encompasses the laws of restricted relativity and the laws of gravitation of Isaac Newton. Hence by theoretically integrating these 4 parameters I came to the conclusion that I will address after the following paragraphs.

In conclusion, the rotary motion, started in the infinitely small (spin), is transmitted in the infinitely large which is made up of the particles of the infinitely small. It is as almost all the mass of an atom is concentrated in a very small positively charged volume : the nucleus; and on the other hand, that the electrons (negatively charged) gravitate around the nucleus like the planets around the Sun. Ultimately, the mass of an atom already possesses a gravitational property acquired during its composition. So gravitation is not the property of mass as magnetite is of the magnet. It is rather a force acquired by the mass during its composition and this under the effect of energy from the original particles.

Thus, in the infinitely large body, the circular movement or rotation produced by the dynamics of giving and receiving between two or more entities, constantly modifies, the angle of revolution around the mother entity (e.g. the sun as the mother entity with its planets) and becomes spherical in the vacuum. Maintained by centripetal and centrifugal force, it rotates around the mother entity creating translation. This can be explained from the Universal Prime Energy which is the fundamental force of God, the Creator. It is this Fundamental Energy that God endowed to creation as the Universal Prime Force and which tends to bring everything back in unison as the dual characteristics in God are in harmony.

Therefore, the force generated within the entity by

its components of the dual characteristics coming from God who is the source, in their action of giving and receiving creates what is called gravitation tending to bring everything back to the center. To say that the driving force of everything is gravity is like saying that boiled water is the basis for the preparation of all food. Boiled water is simply a state or the result of a combination of several actions.

Concerning what we have just said, namely that the gravitation in a body tends to bring everything back to the center, this is confirmed by this scientific sentence :

"In a solid or liquid body, the mutual forces of attraction between all the particles are composed, and everything happens as if the particles were attracted to one and the same point, which is the center of gravity of the body. This is how, on the surface of the Earth, bodies are attracted to the entire mass of the Earth as if it were gathered at the center. This is the reason for gravity, and this is why objects fall vertically towards the center of the planet" (Cf : r : futurascience.com/10essential questions about the Universe/ on 19/08/2019).

In other words, in a body, the magnetic pole at the top tends to unite with the magnetic pole at the bottom, according to Coulomb's law, creating a series of rotations like an animal doing everything it can to catch its tail. This multiplies the force of gravity and creates the difference in the gravitational forces between the poles and the center called the equator. As far as the earth is concerned, we can read :

"The force of gravity on an object is not the same over the entire surface of the Earth. At the equator, the centrifugal force is strong, which relatively reduces the measured gravitational force. On the other hand, the centrifugal force at

the poles is zero, which relatively increases the gravitational force". (Cf : "gravitation." Microsoft® Encarta® 2009 [DVD]. Microsoft Corporation, 2008).

Therefore, gravity, as we have said, is only a form of energy that tends to bring everything back to the center. This is how spheres are formed. It's like spreading iron filings around a magnet. The pieces of iron will move and gather together to form curves.

For illustration, the asteroid RYUGU, small as it is (at least one-kilometer-wide in its largest dimension) does not escape the rule of what we have just said as we can read :

"*The period of rotation of the asteroid RYUGU is about 7 hours and 30 minutes, it is the time it takes to make a turn on its own axis*". (Cf : futurascience.com, Hayabusa 2 : Mascot has landed on Ryugu and is starting its scientific activities, by Remy Decourt, Published on 04/10/2018).

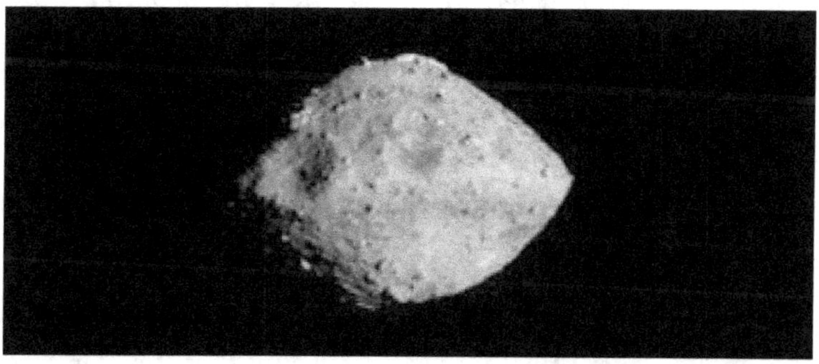

Image : JAXA, University of Tokyo, AIST

All this can be the subject of a book if one wants to explain in detail the combination of all its forces in a single mass.

In fact, mass is the external projection of everything that emanates from the particles. And its gravitational force is a resultant. Hence this conclusion : "What is commonly called the "force of gravity" is in fact a combination of forces". (Cf : "gravitation." Microsoft® Encarta® 2009 [DVD]. Microsoft Corporation, 2008).

Despite this, we cannot perfectly explain the phenomena of the Universe because there will always be an invisible aspect that will escape us. This is the action of the Universal Prime Energy in all phenomena. After the Rev. Dr. Sun Myung Moon goes to lay down the principle, the details are to be discovered by biologists and physicists.

We had said that the explanation of the symbolism (++-) (– -+) within the elements was the greatest discovery that has swept away the enigma for everything concerning body dynamics in the Universe. And, we followed it with this statement by the scientists :

"*Without claiming that our Universe turns round, we have to admit that everywhere, from the closest to the most distant objects, from the modest to the most gigantic, we observe a great unbridled waltz : the proper rotation movements of planets, stars, galaxies, the revolution movements of these same planets, double stars...*". (Cfr :H. GIÉ ; The kinetic moment across the Universe* BULLETIN OF THE UNION OF PHYSICIANS P.427 N°712, Paris).

It is this dynamic of the Universe, observed by scientists, that makes scientifically true everything that Rev. Dr. Sun Myung Moon wrote in the Book of the Divine Principle, concerning the points we have selected as part of the physics of creation. In the same book of the Divine Principle, Rev. Dr. Sun Myung Moon started below the science, through the science, and concluded beyond the science to conclude on the principles of restoration which concerns the theologians and why not the whole of humanity.

Thus, for the Universal Cosmological Model, the driving force of everything is energy, starting at the origin with the Universal Prime Energy that is in God, the basis of his eternity. This energy, according to the Principle of Creation, chapter of the Divine Principle of Rev. Dr. Sun Myung Moon, generates in God the Universal Primal Force; which is the fundamental force of God, the Creator.

It is also the fundamental force of creation, the force with which God endows each being or each individual body with truth when it is created as its substantial object according to the diagram below :

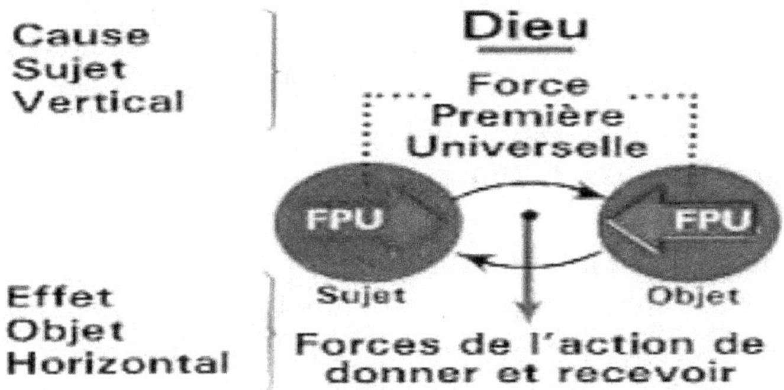

Source : Divine Principle, Level 4, p. 21

In God, there is the Universal Prime Energy, whose combination of all the forces within him on the basis of this Energy generates the Universal Prime Force. This Universal Prime Force, transmitted within the entities, generates relationships for their cohesion. By subject and object, which are religious terms, we must take two entities such as positivity – negativity, energy – mass, spirit – body, man – woman, planet – satellite, etc...

"In other words, thanks to the First Universal Force, the action of giving and receiving is directed by a goal of unification and, thanks to its reciprocal relations (between

subject and object), generates the forces necessary for the existence, reproduction and action of all things, from the smallest to the largest". (Cf : Divine Principle, Level 4, Pp. 21-23).

4.2. The origin of gravity

"*Gravity was born during the Big-bang*" (Stephen HAWKING)

In order for the Big-bang itself to take place, it first required a condensation of the elements described in the Bible as water in the abyss. In the Jerusalem version of the Bible we speak of a wind from God that was able to unite them (Gen 1:2). But in the common language of science, many scientists began to equate it with a super gravity whose gravity broke away causing the Big-bang. Thus, gravity was not born during the Big-bang. Since there was already condensed material or the Universe condensed as the scientists say before the Big-bang, the gravitational force was already present and preceded the Big-bang.

And, as I said in the previous point, Gravitation is the expression of the presence of masses, therefore of material, itself "condensed" energy. In other words, gravity is only a summation of all the inner forces of material stored up from the elementary particle. This notion can be confirmed by the following quotation :

"*Material is indeed produced at the interstices of vacuum bubbles. Gravitation could well be only the product of a sum of pressures of the vacuum bubbles...*"(Cf : http://www.matierevolution.fr/spip.php? article, Les bulles de vide et la matière, le 16/05/2008, Robert Paris).

Section 5 : The prophecy of Stephen HAWKING

5.1. The fulfillment of Stephen Hawking's prophecy

I have always said that, God is working strongly with the scientists on the scientific level for the fulfillment of the third blessing, which is the ability for man to dominate everything in creation (dominate the earth and subdue it). (Cf : Presentation of the Divine Principle ; ed. Culture et Paix, Paris, P. 47).

Since scientists are always occupied with scientific ideas, their thoughts often remain pure. It is thanks to them, that today I can follow a sermon given in Seoul live in Africa. Through their ideas, they sometimes transcend time and space without knowing it and come back to earth with something that at first seemed to be madness. At that moment, God can speak to them without them being aware of it. This is explained in detail in Chapter V of the book of the Divine Principle by Rev. Dr. Sun Myung Moon, a chapter that has now sealed the problem of reincarnation and resurrection that Jesus left unresolved.

Man is a channel for the inspirations of good and evil ; everything depends on his choice. Thus, you cannot imagine my joy when I discovered a divine inspiration captured by Stephen HAWKING and which was fulfilled with the publication of my discoveries that have marked the beginning of a new era in cosmology. It was the advent of the biblical model restored in the Universal Cosmological Model. Let's read this statement by Stephen HAWKING :

"If we really manage to discover a unifying theory, it should in time be comprehensible to everyone in its grand principles, not just a handful of scientists. Philosophers, scientists and ordinary people, all will be able to take part in the discussion about why we exist and why our Universe exists. And if we ever find the answer, it will be the triumph of human reason, which will then allow us to know the thought of God." (Cfr : Stephen Hawking, A Brief History of Time, trans. I. Naddeo-Souriau, Paris, Flammarion, 1989, p. 213).

The advent of this discovery has just been accomplished with the arrival of Astrotheology, which combines the Principles of Creation, a branch of the book of the Divine Principle of Rev. Dr. Sun Myung, where the "why of our existence and of our Universe" is explained, and science. That's why on the cover we mentioned "Cosmology for Dummies". By reading my books, anyone can now understand cosmology and take part in the discussion about the Universe as predicted by Stephen HAWKING.

Thus the Universal Cosmological Model opens the way to the unifying theory as the scheme of the presentation of the Universe from the Before Big-bang to man is presented in a logical way. It does not close the debate, but rather opens it. Thus, like the Standard Model, it can be enriched.

Enrichment in science consists either in ruling out an error, or in completing an idea or theory according to calculations or observations. But, a true researcher cannot declare a model false in its entirety. Only atheists, who never read me, declare the Universal Cosmological Model to be false, simply because for them the Bible is only made up of fairy tales. So to learn that someone has just proved that one of these fairy tales is a reality, is a bomb in their world.

In the scientific conference "*Journey to the Big-bang*", the Frenchman Christophe GALFARD, Dr. in Theoretical

Physics said "there are laws in nature that cannot change. The search for these laws is theoretical physics". Therefore, as long as there is research, there will always be enrichment.

5.2. Stephen Hawking was a believer

A believer is not only one who worships God. An Atheist who fights God is also a believer, for how can one fight a being whose existence one does not believe in? I would rather say that a true atheist is one who doesn't care about God and Satan. But, as soon as one begins to make comments or anti-comments about one, one acknowledges its existence. Even Karl Marx was a believer, even though it is said that religion made him an atheist.

Besides, I have not found any book where the scholar Stephen Hawking has claimed that God does not exist. All I could read in his statement is that "the Universe did not need God to be created. It was only a set of physical phenomena that led to its emergence".

In the Bible it is written : "In the beginning God created the heavens and the earth, and the earth was formless and empty" (Gen 1:1-2).

God being invisible even in creating, it is logical that the one that doesn't see him declares that the things are created by themselves. It is logical also when Gabriele Veneziano has said that the universe is the outcome of an infinite emptiness. But, the believers know that the infinite emptiness was first of all God himself, but, it cannot be demonstrated. This was confirmed by Pope Francis in this quote : "The beginning of the world is not a result of a chaos that owes its origins to another chaos, but derives directly from a supreme Principle who creates out of love."

(Cf : https://fr.aleteia.org/2016/12/08/quand-la-religion-rencontre-la-science).

We have said in our many commentaries that as soon as the word 'earth' was first mentioned in the Bible, all the processes up to the creation of man were physical, with God working in the shadows or in the spirit. It is on this physical aspect that all scholars have based themselves.

God's name is repeatedly mentioned in the debate on cosmology. In response to the Danish physicist Niels Bohr's statement that there is no place for coincidence in creation, in this talk about fundamental indeterminism, whereby the results of measurements depend on chance in a way that is impossible to predict, Albert Einstein replied that "God does not play dice". Stephen Hawking contradicted this by saying : "*Not only does God play dice, but sometimes he throws where we cannot see him.*" (Cfr : J.Baker : 50 clés pour comprendre l'astronomie, ed. Dunod, Paris 2016, p. 99).

Here, Stephen Hawking was surely referring to the dimension of the Universe, whatever the context in which he said this sentence, for us it is a glorification that he brought to God after much reflection. So, for us Stephen Hawking was known and loved by God because of his ideas, which if analyzed closely, glorified God.

Through what scholars have called Hawking's radiance, he partially explained in a scientific way the biblical verse : "...a wind of God moving on the water, ... and the light shone! "(Gen 1:2-4). From this we can conclude that Stephen Hawking was a believer, but whose rational reasoning made him appear to be an atheist.

Chapter XIII :
Science and Religion

Section 1 : Unite between science and religion

1.1 Harmonization between science and religion

Can science and religion unite ? If Adam and Eve hadn't sinned, we wouldn't even be talking about religion. The relationship between man and God would have been normal and natural like the one that unites us to our parents. The two disciplines which seem to be diverging today should rather come to a point of convergence one day as it was originally.

When we speak of religion, we are not talking about worship, which is the only vision that atheists have of believers, but rather about what it teaches. Both religious and scientific teachings seek to dispel ignorance, as we can read in these introductory sentences of the Divine Principle :

"From the dawn of history to the present day, human beings have never ceased to search for the truth that will enable them to overcome both types of ignorance in order to attain knowledge. Humanity has taken the path of religion to seek the inner truth, while pursuing the outer truth through the path of science. Religion and science, each in its own field, have been the methods for seeking truth in order to overcome ignorance and attain knowledge. Ultimately, the path of religion and the path of science should lead to a combined approach to solving their problems; both aspects of truth, inner and outer, should develop in full communion. Only then will we taste eternal happiness, completely free from ignorance and living totally in goodness, in accordance with the desires of our original soul.

(Cf : Presentation of the divine principle : ed. Culture et

Paix, Paris, 2013, Pp29-30).

Source : Divine Principle Level 4, p18

– Tim 2:3-4 *"This is good and acceptable in the sight of God our Savior, who wants all men to be saved and to come to the knowledge of the truth.*

1.1. God inspires those who think about the genesis of the universe

In his address to the plenary session of the Pontifical Academy of Sciences, Pope Pius XII said :

"In truth, the created world is a manifestation of the wisdom and goodness of God, for all things have received their existence from Him and reflect His greatness. Each of them is, so to speak, one of his words, and bears the mark of what we might call the fundamental alphabet... Created things are words of truth. In themselves, in their being, there is neither contradiction nor confusion. On the contrary, they are always coherent with each other". (Cfr : Pius XII : https://inters.org/Pius-XII-Structure-Matter-God. 24 avril 1955).

I said at the beginning that God speaks to scholars without them noticing. Here are a few cases :

– For believers and laymen in astronomy, when science talks about dark energy and dark material, it does not mean the blackness of color, but rather it is an invisible energy and material that science has discovered, and is in action, attributing it even to the participation of the creation of the Universe. But it is also difficult to see what is in the darkness. Moses too, has the knowledge of this energy, called it darkness, that is to say, synonymous with black or invisible. Between dark energy and darkness, we see that the source of inspiration of the names is the same, and this source comes from God.

– It is the same between light and radiance. In speaking of the waves of light emitted by the sun, we say the ray of the sun. Moses used the term "light," but when science discovered the waves of heavenly light, it called it radiation. Between Moses' "light" and the "radiance" of science, we again feel that the source of inspiration for the name is the same, it is God.

– And one more point that astonishes men of science in these revelations is the designation of the word "waters" as the primitive material of the Universe by Moses and in ancient philosophy. After the discovery of hydrogen by Henry Cavendish in 1766, Lavoisier did not even know that hydrogen occupied most of the Universe. Without even referring to the Bible, he called it "hydrogen". Now, hydro means water. In water we have hydrogen and oxygen. This shows again that the name of this element and that of Moses come from the same source ; it is God our Father.

Thus Moses deserves to be called "a scholar of scholars" and the other volumes will not cease to confirm this.

Section 2 :
God is the progenitor of the cosmos

2.1. The Universal Cosmological Model is scientific

The Universal Cosmological Model is scientific and religious. To this day the standard model continues to be enriched. Just as in the future what were believed to be religious theories will begin to reveal their scientific content.

Indeed, in the Bible there is only truth, but in science there are theories. A truth can have limits and be banished depending on the circumstances. It can only be replaced by another truth, as Saint Paul said : "When that which is perfect comes, that which is partial will disappear" (1 Cor 13:10). A theory, on the other hand, is something that can be debated, that can be changed and that can be destroyed at any time. It is the fruit of man's intelligence. And, according to Emile Durkheim, quoted by J. Baker, a law is created after a well-made observation, as well as, often, a well-conducted experiment. (Cf : J.Baker : 50 clés pour comprendre l'astronomie, ed. Dunod, Paris, p.175).

Also, in this book, what is a truth and which is not debatable about the Pre Big-bang, is what we can read about the Bible :

Ge 1:2" *The earth was formless and void : there was darkness over the surface of the deep, and the Spirit of God moved over the waters. 3 God said, "Let there be light. And the light was. 4 God saw that the light was good; and God separated the light from the darkness. 5 God called the light day, and the darkness he called night. So there was evening and there was morning—the first day. »*

The scientific and uncommented version of all these verses, we have named it "the Universal Cosmological Model". In these verses the great discoveries are the following :

– The earth represented the Universe at its beginning where there was nothing. (Gen 1:2)

– Darkness represents different energies including dark energy.

– The abyss represents the space of concentration of the original particles (original material to be better understood by the layman) and one speaks of a hole when in the primitive vacuum, something was formed that has limits. It is the first deformation of the fabric of space-time, even the first black hole.

– The spirit of God or a wind of God represents the first force with gravitational properties that attracts the material Pre Big-bang.

– The waters represent the original particles and original material.

– The light that came at the end, represents radiation.

– The separation of the waters from the waters represents the Big- bang.

I do not close the parenthesis, but rather, I open it, and leave the place to the scientists in their research. If one of the associations here does not fit with its scientific correspondent, the Bible keeping the absolutism of its truth, it is rather up to science to discover to which scientific word it can be associated. But this in no way destroys the Universal Cosmological Model.

I insistently remind researchers that the Universal Cosmological Model is not imposed on them as an act of faith. Let them do the analyses on all my arguments. Moreover, there are many questions that were unanswered in the other Models, but, by drawing on the arguments of the Universal Cosmological Model, the veil is lifted.

To date, natural science has not been able to discover anything about the Universe that can cast doubt on the biblical scheme of creation. The Big-bang had become the point of origin for all calculations on the Universe. But, the Universal Cosmological Model has just moved these calculations back to the Pre Big-bang. It is currently orienting a whole science on a new path with ideals to reach. As a result, all the models are swallowed up in the "Universal Cosmological Model".

Moses and Empedocles, according to my research so far, are the only ones who thought about the origins of the universe before Christ and offered the models. After Jesus Christ, until the 21st century, there was only Rev. Sun Myung Moon who made a new contribution by reasoning on the origins of the inner forces, source of the dynamism of the cosmos. Other scientists think only about the elements of the universe and its functioning. Those who wanted to go to the level of the origins, all failed by dismissing God.

After the Big-bang, humanity lives in a world of results of which we are a part. This does not mean that we will have to deny God because we cannot develop our research beyond the cause. That is why all cosmological models should only be the development of the biblical scheme of creation.

As I said, this sentence is a divine recommendation and insistence of the one who created the cosmos. Whether we believe it or not, it does not change anything, but it does not mean that we should not believe in the truthfulness of the Biblical scheme which is the first cosmological scheme, scientifically restored in the Universal Cosmological Model.

From now on, the Universal Cosmological Model, which

scholars should recognize scientifically, should serve as a basis for other models. Now, with "The Universal Cosmological Model", God has just entered science. It offers a more thorough scheme than others and clarifies many dark points and even corrects the conceptions of some models.

The Standard Model that reigned for a century no longer answers many questions, as Prof. K. K. has pointed out. Aurelian Barreau in his book Big-bang and beyond (ed. DUNOD) : "The Big-bang theory in cosmology is satisfactory in many respects, but does not answer observed anomalies, such as dark energy and dark material".

Now, the Universal Cosmological Model, although it does not explain how dark energy appeared, nevertheless confirms its existence (darkness) before the Big-bang, as well as the presence of another not insignificant force (A wind of God) and another physical manifestation of size which still remains an enigma despite my explanations.

This enigma is contained in the verse "God separated the light from the darkness". (Gen 1:4). I had said that the name of God intervenes only to demonstrate His action, His will and His plan, but the phenomenon itself was physical. It was through this that I suspected the presence of dark material, responsible for this separation.

Now, with the series of books "The Verses of the Scholars", God has just entered the scientific debate as the Creator, all this to make the scientists say that the "Universal Cosmological Model" is a scientific model.

2.2. The universe is a complex machine

Complex means "*composed of many elements which form a whole difficult model to apprehend*".(Cf : Dicosencarta

2009)

So whether we like it or not, the universe is more complicated than the most complicated of Earth's engines. Without spatio-temporal knowledge like that of Moses, we will not know how to link the joints and the models will always remain incomplete. the individual knowledge for each one on the universe is worth only one for infinity. Over generations, knowledge will always be assembled and this constitutes the advancement of cosmology. Who would have believed that after 3,500 years of deadlock, the history of biblical creation will be linked to modern science, resuscitating Empedocles and bringing new horizons of research? The 100% of knowledge is held only by God alone.

General conclusion

I insistently remind researchers that the Universal Cosmetic Model is not imposed on them as an act of faith. Let them analyze all of my arguments. Besides, several questions that were unanswered in the other Models, but, drawing on the arguments of the Universal Cosmological Model, the veil is lifted.

For biologists, there is nothing to comment on the origin of life. It all started from the elements found in the Universe that the Bible and the Qur'an refer to as the dust of the earth. The Bible being always a living word, The Universal Cosmological Model brings more knowledge in cosmology. Based only on the main themes, this book is universal to have :

– Scientifically reconstruct the history of Biblical creation

– Talk about the Pre-Big Bang with a lot of confidence;

– Claimed that dark energy and dark material are leftovers from before the Big Bang

– Discovered that the Big Bang is biblical, this comforts believers;

– State that the fossil radiation is also biblical and, appeared in the boundary between the Pre Big-bang and the Big-bang;

– To state that the Bible in speaking of the development of plants on the third day, that is, at the time of galactic collisions, meant that the molecules of life were found in cosmic dust. But, what is found on the planet earth carried life.

– To state on the basis of biblical verses that the winged beings appeared after the fish, scientifically confirmed by the fossils of the dinosaurs;

– Demonstrate that man is made of the dust of the earth, which is an extension of the star dust from the condensed universe, created by God.

I have already said that the cosmos in general and the universe in particular is a complete body like a man. The fact that it may have less than one model to explain it is abnormal, because it only took existence from one. So, we don't reject any model, but, the sooner we encompass them, correct and turn them into theory and this confirms the Universal Cosmological Model.

Now, with the series of books "The Verses of the scholars", God has just entered the scientific debate as the Creator.

BIBLIOGRAPHY

- SCIENTIFIC LITERATURE

– Barrau, Aurélien : *Big-bang et au-delà*, balade en cosmologie, aux éditions Dunod, Paris.

– Basset, Bruce & Edney Ralph, *Relativité*, Icon books ltd, Londres, 2009.

– Bender, Lionel : *les plantes*, éd. Artis-Historia, Bruxelles, 1989.

– Sarrailh, Claude, *Le secret de la vie*, éd. Librinova, 2018

– Douce Roland & Postaire Eric, *Les origines de la vie*, Gallimard, 2016.

– Galfard, Christophe, *E=mc*, Flammarion, 2017

– Igor & Bogdanov Grichka, *Avant le Big-bang*, Grasset, 2004.

– Baker, Joanne : *50 clés pour comprendre l'astronomie*, éd. Dunod, 2016.

– Herrmann, Joachim, *Atlas d'Astronomie*, ed. Stock, Paris, 1976

– Klein, Etienne *: Les secrets de la matière*, éd. J'ai lu, Paris, 2018.

– Klein, Etienne : Discours sur m'origine de l'univers, éd. Flammarion 2016.

– Pluchet, Blandine, *La physique quantique pour les nuls*, éd. First, 2018.

– Reeves, Hubert : *L'heure de l'ivresse; L'univers a-t-il un sens ?* Le Seuil, 1986

– Sagan, Carl : *Cosmic Connection*, ed. du Seuil, Paris, 1975

– Whyntie & Pugh, Oliver, *La physique des particules en image*, EDP Sciences, 2014.

-- Heidmann, jean : Sommes-nous seuls dans l'univers, Fayard, Paris, 2000.

– RELIGIOUS BOOKS

– Louis Segond Bible
– Jerusalem Bible
– BABYLON THE GREAT IS FALLEN - the kingdom of God begins its reign, Watchtower, JW.org,
- *Présentation du Principe divin* : éd. Culture et Paix, Paris, 2013
– *Le Principe divin*, niveau 4, introduction
– FF¨PMU : The thought of unification
– VIE UNIVERSELLE : Les champs de rayonnement, Würzburg 1, 1986.

– COLLECTIONS

– Les défis de la science : l'origine de la vie, édité par RBA coleccionables, S.A.U. 2018, Barcelone.
– Les défis de la science : La théorie du Tout, édité par RBA coleccionables, S.A.U. 2019, Barcelone.
– Les défis de la science : Au-delà du Boson de Higgs, édité par RBA coleccionables, S.A.U. 2018, Barcelone.
– Voyage dans le cosmos : Les trous noirs, édité par RBA coleccionables, S.A.U. 2020, Barcelone.
– Voyage dans le cosmos : Les exoplanètes, édité par RBA coleccionables, S.A.U. 2020, Barcelone.
– Voyage dans le cosmos : Le boson de Higgs, édité par RBA coleccionables, S.A.U. 2020, Barcelone.
– Voyage dans le cosmos : L'espace-temps quantique, édité par RBA coleccionables, S.A.U. 2020, Barcelone.
– Voyage dans le cosmos : Les univers parallèles, édité par RBA coleccionables, S.A.U. 2020, Barcelone.

– Voyage dans le cosmos : Les trous noirs, édité par RBA coleccionables, S.A.U. 2020, Barcelone.
– Voyage dans le cosmos : La matière noire, édité par RBA coleccionables, S.A.U. 2020, Barcelone.
– Voyage dans le cosmos : Le vide et le néant, édité par RBA coleccionables, S.A.U. 2020, Barcelone.

– COURSE & PRACTICAL WORK

– Lachieze-Rey, Marc : Initiation à la comologie, 5è éd, Dunot, Paris, 2013.
– Matthews, P.T., Introduction à la mécanique quantique, Dunod, Paris, 1970.
– MPANZU K.B. and NIMI M., explanation of quantum gravity, National Pedagogical University (RDC), 2017, unpublished

– REVIEW

– Science & Vie Special issue n ° 288, 09/2019,
– GEO HORS-SERIE, Why do we need trees so much, August – September 2019
– Les secrets de l'espace 07 : # HORS SERIE#, Les mystères de l'Univers.

– ENCYCLOPEDIA

– De l' Astronomie de l'Espace : Les forces fondamentales, Hachette, 2019.
– De l' Astronomie de l'Espace : Les trous noirs, Hachette, 2018
– De l' Astronomie de l'Espace : Matière noire et énergie sombre, Hachette, 2019.
– Edey, Maitland A., Les origins de l'Homme ; Le Chaînon manquant, éd. France Loisirs, Paris, 1984

– Geographic exploration. "Microsoft® Encarta® 2009 [DVD]. Microsoft Corporation, 2008
– Grint, Damian, *Les mystères de l'univers*, éd. Christophe Colomb, Paris, 1985
– Joe Bergeron, Encarta ® 2009 Modeling a black hole
– kid, Linda, *Les origines de la vie*, éd. Artis-Historia, Bruxelles. 1989
– Melvin & G. Berger : *Qu'est-ce qui fait onduler l'océan ?* ed. Nouveaux Horizons - ARS, Paris 2012
– Moïse. " Microsoft® Encarta® 2009 [DVD]. Microsoft Corporation, 2008
– P. Batchelor : *L'énergie, le son et la lumière*, éd. Nouveaux Horizons, ARS, Paris, 2016
– Wikipédia ; *Particules élémentaires*
– Wikipedia ; Metaphysical
– Zoppi, Moninga, Les Merveilles de l'univers, Médiaspaul, Kinshasa, 2009.

– WEBSITES

– Anne-Laure Lebrun, Figaro.fr, published on 02/05/2018
– annuairediggdes.coursgratuits.net/c7_bigbang/p715_part virtuelles. phpparOlivier Esslinger
– astronomes.com/le-big-bang/particule-virtuelle/ August 9, 2017 by Oli-vierEsslinger
– Astrophysical Journal of June 10, 2015) " : the song of the stars, the new stellar music
– At the origin of the Big Bang : Abbé Georges Lemaître (Humanity ; Science and religion)
– Didier VerkinDe26 / 06/1999 : http : //wwwlapp.in2p3.fr/neutrinos/nsources.html
– Dr. David P. Stern Mail to Dr. Stern : education ("at" symbol) phy6.or
– F. J. Dyson, Infinite in all Directions, Harper and Row, 1988

– futurasciences.com. First stars in the universe : signals finally detected, Laurent SACCO. The 02/03/2018
– H. GIÉ : Angular momentum through the Universe * BULLETIN DE L'UNION DES PHYSICIENS P.427 N ° 712, Paris
– https://ingeniumcanada.org/innovation/fr/recit/un-trou -noir-l origin- de-lunivers.
– lemonde.fr, What the work of Stephen Hawking brought to physics, by Gary Dagorn, on 03.14.2018
– Mark Eichenlaub in Quora : What did Richard Feynman mean when he said, "What I cannot create, I do not understand" ?
– matierevolution.fr, Dark matter, dark energy : the missing link ? 11/13/2007, by FaberSperber, Robert Paris
– monde-diplomatique.fr/2012/07/VOILLIOT/47999 - summary taken from the book "Man according to Karl Marx by YvonQuinioun
– Quora.com,
– scienceetavenir.fr ; What is the temperature of the Universe ? by Erwan LECOMTE, the 09/28/2018
– Jean-Luc Goudet, Futurascience, The 09/10/2014 at 11:38
– The force that comes from the void - Astrid Lambrecht in monthly n ° 376 /06/2004, p.4
– trustmyscience.com/telescope-fermi-retrace-histoire-rayonnement- stellaire-dans-univers/by Thomas Boisson, 11/30/2018
– vinsvignesvignerons.com/Geologie/Une-breve-histoire -de-la- planete-terre
– www.techno-science.net ; Successful test of Einstein's theory of relativity near a supermassive black hole by Adrien "07/27/2018
– zevengeur.wordpress.com/2012/12/sphc3a8re-dunivers-observable-03.jpg
– : fr.sputniknews.com : a black hole emits mysterious

sounds

– SCIENTIFIC DOCUMENTARIES

– "Discourse on the origin of the universe", Etienne-Klein – 2010)

– "HAS THE UNIVERSE EVER BEGINNING ? "Conference by ETIENNE KLEIN

– "New Theory of the Unified Field (Explanation of Gravitation)

– "The end of the universe

– "The mysteries of the Universe beyond the cold", written by Colin CAMPBELL

– "The Origins of the Universe" by Prof. Dr. A.E. WILDER Smith

– "The Unexplained Phenomena of the Universe" written by Laura Verklan Armstrong

– A baby at all costs "on ARTE. (Cfr : arte.tv/fr/videos/079475-003- A/arte-regards/a baby at all costs

– Aurélien Barrau (Cfr : conference : What was there before the Big- bang ? ".

– Cosmos - Beyond the Big-bang »written by Matt HICKEY

– Dr. FredericSaldmann : "Any death before 120 is a premature death in TEDxMarseille

– Hubble views from space : the work of light

– New theory of the unified field : explanation of the dynamic universe

– On the origins of the universe

– Quantum physics for dummies

– Secrets of the Cosmos : The Higgs Field Theory.

– The Dark Ideas of Physics

– The keys to the universe 4 - The origin of water on earth

– The Milky Way and its huge black hole ", by director Thomas LUCAS.

– The Mysteries of the Universe
– The origins of the earth"; written by Alice HARPER and Nigel HENBEST

Big-Bang
TO
HUMANKIND

The Volume I of this book revealing the sphere of the Before Big Bang and its structures was published in the collection "verses of scholars" under the title "Cosmology according to biblical and ancient models". "From Big-bang to Humankind" of volume III in Astrotheology constitutes the continuity.

If the greatest discovery of the last century in cosmology was "The Big Bang" on which all indisputable evidence converges, 100 years later, the greatest revelation is that this phenomenon is also biblical. The author of the book, Amateur astronomer, independent writer and researcher, is the first to have completely restored the biblical creation scheme in its true context, demonstrating that it is the most complete scheme of which science has not yet discovered all the mysteries. The restitution of this scheme brings a new invention in cosmology, namely "The Universal Cosmological Model". It is now up to science to take hold of it and to enlarge it.

At the end of this book, a demonstration of Astrotheology on ancient Chinese and Western philosophies, reveals to us that the models of the Universe have always been consistent even in cosmogony for informed readers.

With a language easy to understand, this book is for everyone...

FLAVIEN PHANZU MWAKA
Born in Kinshasa, June 11, 1967
Congolese nationality,
Amateur astronomer
Independent researcher,
Inventor of the Universal Cosmological Model and Astrotheology

Sommaire

Dépôt légal février 2022

Librinova

www.librinova.com

www.ingramcontent.com/pod-product-compliance
Lightning Source LLC
LaVergne TN
LVHW052257200926
848481LV00002B/394